PIRATES
AND THE
LOST TEMPLAR FLEET

Adventures Unlimited Press

PIRATES
AND THE
LOST TEMPLAR FLEET

The Secret Naval War
Between the
Knights Templar
and the Vatican

by
David Hatcher Childress

PIRATES AND THE LOST TEMPLAR FLEET

ISBN 1-931882-18-5

Printed in the United States of America

Published by
Adventures Unlimited Press
One Adventure Place
Kempton, Illinois 60946 USA
auphq@frontiernet.net

www.wexclub.com/aup
www.adventuresunlimitedpress.com
www.adventuresunlimited.nl

10 9 8 7 6 5

PIRATES
AND THE
LOST TEMPLAR FLEET

The Secret Naval War Between the Knights Templar and the Vatican

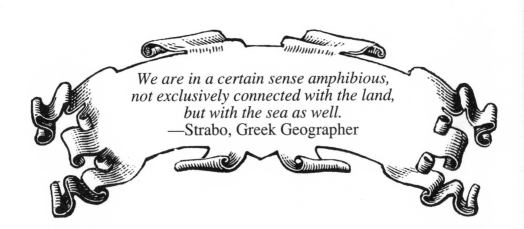

We are in a certain sense amphibious,
not exclusively connected with the land,
but with the sea as well.
—Strabo, Greek Geographer

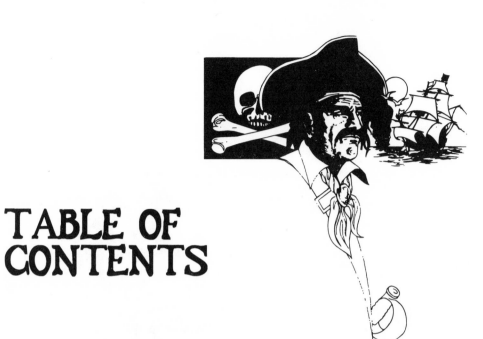

TABLE OF CONTENTS

Dedicated to Pirates Everywhere

1.
10,000 Years
of Seafaring and Piracy

You never had time to learn.
They threw you in and told you the rules
and the first time they caught you off base they killed you..
—Ernest Hemingway, *A Farewell to Arms*

The myth of the noble savage is bull.
People are born to survive.
—Sam Peckinpah

A Pirate's Life For Me...

When I was younger, I wore an eyepatch, carried a wooden sword, and swore at the scurvy dogs around me. It was a pirate's life for me, and until my mom called me in for dinner there was plunder to be had and treasure to be found. I thought of pirates as Robin Hood or Zorro characters—outlaws but honorable among the common people.

As I grew older, my former life as a pirate faded behind me, but as I researched history and archeology for my *Lost Cities* books, I kept running across my former pirate compatriots, men of daring and world travel. I also became interested in the Knights Templar, secret societies and pre-Columbian voyages to the Americas. When I discovered that much of the modern lore of pirates derived from tales of the lost fleet of the Knights Templar, I was thrilled to find this link between some of my favorite subjects. I determined to investigate further.

Any discussion of pirates should start at the beginning of piracy, but when would this have been? As we shall see, there is a lot of evidence that it was very early indeed. In fact, one could speculate that piracy goes back to the very beginning of navigation and seafaring itself.

The chronology of the development and use of seafaring skills is also is a subject of great speculation. We are commonly taught by mainstream historians that although primitive man may have had some boats and fishing capabilities, the development of sophisticated seafaring techniques came at a much later time. This point of view casts our early ancestors as unimaginative monkeymen who were afraid of the ocean. But much of how we reconstruct the past is based on how we perceive the past and the human beings who lived this bygone period. If we instead believe that early men were curious and inventive explorers, the idea that they would ignore the seas would be ridiculous. In my opinion, mankind took to the seas many thousands of years before commonly thought.

Since over two-thirds of planet Earth is covered with water, the ability to exploit water with flotation devices such as rafts, canoes, boats and ships, would be advantageous.

The highways of the ancient world were the seas, rivers and lakes. Floating down a river toward a port city located where the river meets the sea or a lake would be much easier than walking the entire distance. Travelling by boat along the coast of some land or the edge of a large lake, would no doubt be swifter and safer then walking through potentially hostile villages or stumbling into a raiding party. Once it is established that travel by water is efficient, what is to keep men from sailing the high seas in search of food and commerce?

§§§

Evidence of Ancient Transoceanic Traffic

Modern historians often maintain that the continents were populated by migrating tribes who walked everywhere. The Americas, Australia, and even islands in Indonesia are said to have been colonized by people walking over land bridges. But why should we limit our explorations of the past by requiring our ancestors to cross oceans using land bridges?

Historians and archeologists consider seafaring as one of the earliest benchmarks in civilization and the date for the beginning of this important technology is continually being pushed back in time. Suffice it to say that ocean-going seafaring as we know it today has been going on for well over 6000 years!

In fact, archeologists at the University of Australia claim that evidence of advanced seafaring is directly connected with the finds of human bones dated as 30,000 years before present found in an underwater cave in the Pacific Island of New Ireland, to the east of Papua New Guinea. Archeologists at the University of Australia say that these humans must have arrived on New Ireland by boat, and could not have walked across any known land bridges.[20]

There is incontrovertible evidence that the Polynesians and Micronesians navigated vast expanses of oceans in outrigger canoes, sailing more than three times the distance between Africa and South America, peopling the islands of the Pacific. If the Polynesians and other cultures could sail vast oceanic distances thousands of years ago, why do modern historians persist in teaching that Christopher Columbus was the first to cross the Atlantic?

Most people have heard speculation that the Vikings sailed their long ships to Greenland and Labrador about a thousand years ago, and the accomplishments of Lief Erikson are gaining acceptance. But can we take seriously more radical proposals of Irish monks sailing up and down North America almost 2000 years ago, as well as Portuguese and Basque fishermen, Roman, Greek and Phoenician explorers, Hebrew gold miners, and Egyptian traders?

Impossible? Why? Is the Atlantic so impassable? Hardly. People have now crossed the Atlantic in rowboats, kayaks, and simple rafts. The ancient sailors of the Mediterranean sailed ships far superior to those in which Columbus sailed across the Atlantic. Why think they would not have undertaken a transoceanic voyage? In fact, on Columbus' second voyage to the New World, he wrote about finding the wreckage of a European ship on the Island of Guadaloupe in the French West Indies.[49]

Many historians find the evidence of visitation to the Americas by ancient explorers and traders overwhelming. In 1976, a Brazilian diver named Jose Roberto Teixeira was spearfishing around a rock off Ilha de Gobernador in the Baia de Guananbara near Rio de Janeiro, when he found three intact Roman amphorae (clay vessels used to hold wine), in an area with several shipwrecks, some dating from the sixteenth century AD. He reported that the area of his find is littered with pottery shards and large pieces of other amphorae.

The Brazilian Institute of Archaeology was extremely interested in these amphorae and sent photos to the Smithsonian Institute, which identified them as Roman. Later, Dr. Elizabeth Lyding Will of the Department of Classics at the University of Massachusetts, Amherst, identified the amphorae as Second to First Century BC, "...apparently manufactured at Kouass, the ancient port of Zilis (Dchar Jedid) on the Atlantic coast of Morocco, southwest of Tangiers." Dr. Michel Ponsich, the archaeologist who had conducted excavations at Kouass, agrees with Dr. Will on the place of manufacture, and gives the amphora a date of Second Century BC.[49]

An American archaeologist who specializes in underwater digs, Robert Marx, investigating the site near Rio de Janeiro where the amphorae were found, located a wooden structure in the muddy bottom of the bay. Using sonar, Marx discovered that there were actually two wrecks at the site, one a sixteenth century ship, and another which was presumably a more ancient ship, the source of the amphorae.

But before Marx could dive to the site, trouble started. Brazilian authorities did not savor the idea of a Roman shipwreck off their coast, and Spain and Portugal are still disputing who first discovered Brazil. Marx was even accused of being an Italian agent sent out to drum up publicity for Rome! Under pressure, the Brazilian authorities refused to grant permission for Marx to keep diving; and later permanently banned him from entering Brazil.

Marx felt that the ship might have been blown off course in a storm. Wrecks thought to be Roman have been found off the Azores in the middle of the Atlantic. As we shall see, plenty of contact with the Americas first occurred by accident. In the last century alone, over 600 forced crossings of the Atlantic have occurred as ships and rafts were blown to the Americas by storms. Yet, I do not personally believe that the Romans were accidentally showing up to sun-bathe on the Copacabana Beach near Rio de Janeiro. More than likely, they were deliberately sailing to the New World!

Many other Roman artifacts have been found in Latin America. A large hoard of Roman jewelry was found in graves near Mexico City by Dr. Garcia Payón of the University of Jalapa in 1961. Roman fibula (clips used to hold together Roman togas), as well as Roman coins have frequently been found. In fact, a ceramic jar containing several hundred Roman coins, bearing dates ranging from the reign of Augustus down to 350 AD and every intervening period, was found on a beach in Venezuela. This cache is now in the Smithsonian Institution. Experts there have stated that the coins are not a misplaced collection belonging to an ancient numismatist, but probably a Roman sailor's ready cash, either concealed in the sand or washed ashore from a shipwreck.[49]

The Romans are not as often associated with world travel as another great ancient power: their deadly rivals, the Phoenicians. In the first century BC, Greek geographer Strabo wrote, "...far famed are the voyages of the Phoenicians, who, a short time after the Trojan War, explored the regions beyond the Pillars of Hercules, founding cities there and in the central Libyan [African] seaboard. Once, while exploring the coast along the shore of Libya, they were driven by strong winds for a great distance out into the ocean. After being tossed for many days, they were carried ashore on an island of considerable size, situated at a great distance to the west of Libya."[49, 57]

The Phoenicians were also known as Carthaginians. The names stemming from the names of their two main colonies, Phoenicia and Carthage. There is an ancient passage detailing their forays into the Atlantic after the end of the Trojan War, circa 1200 BC.

But how far did they go?

In 1872, near Paraiba, Brazil, a stone bearing a Phoenician inscription was discovered. It was thought to be a forgery for almost a century, when in 1968, Dr. Cyrus Gordon, Chairman of the Department of Mediterranean Studies at Brandeis University, announced that the inscription was genuine. Copies of the Paraiba inscription speak of a Phoenician ship circumnavigating Africa until it was blown to the shores of Brazil.[57, 92] Indeed, the moderm "discoverer" of Brazil, Portuguese explorer Pedro Alvares Cabral, was attempting to round Africa in AD 1500 when he was blown off course and landed in South America. It is believed that he named Brazil after after the legendary Irish island of Hy-Brazil.

A Carthaginian shipwreck containing a cargo of amphorae was discovered in 1972 off the coast of Honduras, according to Dr. Elizabeth Will.[49] Had regular trade routes opened up? Some scholars believe that the Toltec Indians were in fact Carthaginians, who, after being defeated by Rome in the Punic Wars, left the Mediterranean for West Africa. From there they migrated to the Yucatan Peninsula of Mexico, where they reestablished their civilization. The Aztecs later destroyed them, and Carthaginian gold bars fell into the hands of the Aztecs, later surfacing in the United States as part of Montezuma's gold and the "Seven Gold Cities of Cibola."

Even the Jews may have reached the Americas. Inscriptions in archaic Hebrew near Las Lunas, New Mexico, supposedly tell of their voyage and the establishment of their city. However, there is a great deal of debate about what the inscription actually says, even by those who believe it genuine. As late as 734 AD, Jews seeking refuge from persecution sailed from Rome "...to Calalus, an unknown land."[16, 109]

If the ardent celebrator of Columbus Day finds this a bit beyond his or her paradigm, there's more. Just a few years earlier, around 725 AD, seven bishops and a reported 5000 followers fleeing from the Moors in Spain sailed from Porto Cale, Portugal, for the island of Antillia. They landed on the west coast of Florida, according to some historians, and made their way inland to found the new city of Cale, which later may have become modern Ocala. The Jews from Rome could have learned of this Portuguese exodus, gone to Porto Cale, hoping for exact sailing directions. Once in America, they called the new land Calalus, a sort of latinized Cale.

The Portuguese voyage was certainly known to Columbus, who thought he would find their decendants on an island. Perhaps he thought that the European wreck he found on his second voyage was from this expedition.[109]

But even earlier Jews may have found their way across the oceans. King Solomon's treasures were found in the mysterious land of Ophir. The locating of this gold-rich country has been the subject of much speculation. Solomon was the son-in-law of the Phoenician King Hiram. With his help, Solomon undertook the amassing of the vast treasure required to build the Temple in Jerusalem.

In the Bible, we are told:

> King Solomon made a fleet of ships in Ezion-Geber, which is beside Eilat on the shore of the Red Sea in the land of Edom. And Hiram sent in the navy of his servants, shipmen that had knowledge of the sea, with the servants of Solomon. And they came to Ophir, and fetched from thence gold, four hundred and twenty talents, and brought it to King Solomon.... Once in three years the fleet came in bringing gold, silver, ivory, apes, peacocks... a very great amount of red sandalwood and precious stones.
>
> *I Kings 9:26-28, 10:22, 11.*

What these great seafarers brought back to Ezion-Geber was a literal fortune in gold: four hundred and fifty talents is almost twenty tons, or 40,000 pounds! The source of that much gold would lure many an adventurer, even today, to cross an ocean or two. But they also brought back silver, ivory, apes, peacocks and more. Yet, the land of Ophir, where all this wealth came from has eluded historians for centuries.

One of the reasons it is so elusive may be that most historians limit their search in the belief that ancient man's seafaring capabilities were limited.

The myopic attitude of Biblical scholars and historians may be summed up by this statement from Manfred Barthel, the German scholar who wrote *What the Bible Really Says*: "Zimbabwe was too recent, India too far away, the Urals too cold...It does seem likely that Ophir was somewhere on the shores of the Red Sea."[89]

Similarly, the more openminded German scholars Hermann and Georg Schreiber say in their book *Vanished Cities*: "At one time the idea arose that the Ophir of the Bible may have been in what is now called Peru. But that is out of the question; no merchant fleet sailed so far in the tenth century B.C. ...The search for the famous land of gold has therefore been restricted fundamentally to southern Arabia and the African coast.[71]

The Schreibers make many good points in their book, but at the same time they fall into the strange logic of the isolationist: they believe that ancient man never ventured far from land or the known, so they ignore the very evidence that might convince them that he did!. Manfred Barthel is typical of the isolationist at his worst. For Barthel, even India is too far away. Even though it took three years (that's right, three years!) to go to Ophir and back, sophisticated navies cannot go from Eilat to India!

When one puts this type of anthropological myopia behind him, Peru starts to look like a possibility. But the other items brought back from Ophir—apes, ivory, peacocks and sandlewood—lead us to look in the other direction. While the apes and ivory have typically led researchers to suspect ports in Africa, both are found in India and Southeast Asia, and more importantly, peacocks especially, and sandlewood incidentally, come from India and Southeast Asia.

There is one location for Ophir that has been overlooked by everyone—Australia. Australia is one of the most mineral-rich countries in the world, and it would be reached by sailing via India and Sumatra. There have been discoveries of ancient mines in both Western and Northern Australia. At the town of Gympie in Queensland, an alleged pyramid was found, now destroyed, along with a three-foot-high statue of the Egyptian god Thoth in the form of a baboon, along with numerous Egyptian and Phoenician relics.[91] What is more, the same place became known as "The Town that Saved Queensland" because of a gold rush there in the late 1800's.

Personally, I think that Solomon's fleet visited ancient mining sites along the coast and rivers of Australia. On the return trip, after growing food for the return voyage, they would stop at ports in Indonesia and India and gather peacocks, apes, ivory and sandlewood.

Egyptian relics in Australia were mentioned a couple of paragraphs back. Is this some mistake?

Curiously, the amazing Egyptian civilization, much older than the Phoenicians, were said to not be an ocean-going civilization. Though their country had coasts on two important bodies of water—the Mediterranean and the Red Sea—and they clearly navigated these routes, they supposedly did not ever go on oceanic exploratory voyages.

Says Marie Parsons in her article "Ships and Boats of Egypt," "There is evidence that the Old Kingdom of Egypt had the first planked boats ever made. These were used in burial rituals. Fourteen have recently been found buried in the region of Abydos... The earliest surviving example of a sewn boat... was found beside the great pyramid of Giza. It is probably a descendant of boats going back into Egypt's predynastic times."

It appears to be Ms. Parsons idea that this boat was fit only for plying the Nile, but this discovery of the huge funerary boat in a stone pit near the Great Pyramid in the 1960s may have given the world a perfectly preserved ocean-going Egyptian vessel that could have sailed around the world. This vessel also has scraping damage on its forward underside that ship experts have claimed could only have been caused by a coral reef.

What is particularly intriguing about the Egyptians vis a vis ocean travel is the persistence of tales of voyages to Punt, an exotic locale from which they imported unguents, dwarves and even giraffes. These voyages go back at least as far as the 5th dynasty. Queen Hapsepsut's 18th dynasty voyages is memorialized in a series of friezes on her temple at Deir el-Bahari.

Again, most historians, based on the "fact" that the Egyptians didn't travel very far, believe that Punt was a nearby African kingdom—pershaps Somalia. But some of the treasures from Punt, such as pygmies, would have had to have come from atleast as far away as Central Africa. This would have required exiting the Red

Sea and travelling down the east coast of Africa to at least as far as Kenya and Tanzania, if not further to Mozambique and Zimbabwe. As we seen, Egyptian artifacts have been found as far afield as Australia and the Pacific, and Mexico in the Atlantic.

Another prehistoric ocean-going civilization was apparently the Hittites, who had naval bases in current-day Lebanon and Cyprus. Evidence exists that they crossed the North Atlantic via the Shetland Island (the "Set-Lands" of ancient Egyptian lore) into the Great Lakes of North America. There they exploited the raw copper at Isle Royale on Lake Superior, and shipped it back to the Mediterranean.[40]

A recent book by the British historian Gavin Menzies claims that ships of the Chinese navy circumnavigated the world in 1421 AD. Menzies claims that the Chinese sent a number of exploratory ships into the Indian Ocean, Atlantic and Pacific, culminating in the voyage of Admiral Zhou Man. Man's voyage in 1421 took him to India, down the east coast of Africa, across the Atlantic, around the bottom of South America and across the Pacific back to China.[59]

Chicago lawyer and historian Henrietta Mertz proposed that the Chinese had made several voyages to the west coast of America in her 1953 book, *Pale Ink*.[82] Oxford professor Louise Levathes details Chinese voyages to Australia and Africa in her book *When China Ruled the Seas*.[61]

The point is that crossing oceans is not very difficult—not now, not in ancient times. Not only could the Portuguese and Christopher Columbus have crossed the Atlantic in the Middle Ages, but just about anyone in an earlier time with a seaworthy ship could have done so as well. In fact, we have the famous Piri Reis map at the Topkapi Museum in Istanbul to testify that the Atlantic was being navigated long before the European Dark Ages "experts" declared that the world was flat. The Piri Reis map shows the entirety of the west coast of North and South America, plus a good portion of Antarctica, and was drafted just a few years after Columbus made his first voyage to the New World. Since it was put together from older maps, some of these presumably in the hands of the Portuguese, Columbus himself probably carried an earlier copy with him. More on these amazing charts in a later chapter.

§§§

Piracy in the Ancient World

As I have noted, sophisticated navigation probably existed far longer than existing documentation indicates, and piracy probably developed along with it, but according to most encyclopedias and dictionaries, piracy dates from the time of the Phoenicians, said to be the world's first seafaring people. The Phoenicians virtually monopolized seatrade in the Mediterranean circa 1000 BC, and were known to attack other ships that were not part of their special network. Theirs were large, powerful ships rowed by scores of men with rams just below sea level for puncturing enemy ships. Indeed, by Phoenician times, piracy and naval warfare had evolved considerably.

The early Romans were tormented by pirates from Cilicia in Anatolia. Three major expeditions were launched (102-67 BC) against the plunderers in the eastern Mediterranean, culminating in their destruction by a large Roman fleet led by Pompey the Great. For the next 250 years, the Roman navy kept the Mediterranean virtually free of pirates.

In his book *Piracy in the Ancient World,* University of Liverpool professor Henry Ormerod says the origin of the English word pirate derives from the Latin, pirata, which means one who makes attempts or attacks on ships.[80]

Ormerod says that piracy in ancient times was mainly situated in the Mediterranean. This area was then, as it is today, a crossroads of shipping traffic. Ormerod points out that much of the coast of the Mediterranean is rocky and barren, unable to support a large population. Says Ormerod, "by land, the poverty of the soil had forced them to become hunters and brigands rather than agriculturalists; the same pursuits were followed on the sea."(80)

Ancient Greek writers give us some incidents of ancient piracy. For instance, in 355 BC, three Athenian ambassadors, who were sailing in a warship to the court of Maussolus in Caria, fell in with a vessel from Naucratis, which they captured and brought to the Athenian port of Pireus. The Naucratite merchants appealed to the citizens of Athens, but since Egypt (where the Naucratites were from) was in revolt from Persia at the time, and the Athenians were anxious to cultivate good relations with the Geat King of Persia, the ship was condemmed as the enemy.

Says Ormerod, "The prize-money, which by law belonged to the state, had been retained by the three ambassadors."

Reprisals were also a part of piracy in ancient times, with ports and city-states being fair game for pirates and privateers to attack and pillage.

Says Ormerod, "It is equally difficult to apply the modern conception of the 'politically organised society' to early conditions of ancient life. It was only as the result of a long process of development that the ancient world came to distinguish between foreigner and enemy, piracy and privateering, lawful trade and kidnapping. To the Roman representations regarding the piracies carried out by her subjects, Queen Teuta replied that it was not the habit of their kings to interfere with the normal pursuits of the Illyrians at sea. Even in sixth century Greece we find Polycrates of Samos, according to Herodotus, carrying on a piracy business directed against all users of the Aegean. A reputed law of Solon seems to have recognized similar proceedings among the Athenians. The plundering of neighbors was to the primitive inhabitant of the Mediterranean area a form of production, which was sanctioned and encouraged by the community, so long as it was directed against the people of a different tribe."

The best description of such conditions says Ormerod is that given by Thucydides:

> For the Grecians in old time, and of the barbarians both those on the continent who lived near the sea, and all who inhabited islands, after they began to cross over more commonly to one another in ships, turned to piracy, under the conduct of their most powerful men, with a view both to their own gain, and to maintenance for the needy, and falling upon towns that were unfortified, and inhabited like villages, they rifled them, and made most of their livelihood by this means; as this employment did not yet involve any disgrace, but rather brought with it somewhat of glory. This

is shown by some that dwell on the continent even at the present day, with whom it is an honor to perform this cleverly; and by the ancient poets, who introduce men asking the question of such as sail to their coasts, in all cases alike, whether they are pirates : as though neither those of whom they inquire, disowned the employment; nor those who were interested in knowing, reproached them with it. They also robbed one another on the continent; and to this day many of the Greeks live after the old fashion; as the Locri Ozolae, the Aetolians and Acarnanians, and those in that part of the continent. And the fashion of wearing arms has continued amongst these continental states from their old trade of piracy.

He also mentions that piracy and brigandage are here regarded as a means of production, and were so classified by Aristotle: "Others support themselves by hunting, which is of different kinds. Some, for example, are brigands, others, who dwell by lakes or marshes or rivers or a sea in which there are fish, are fishermen and others live by the pursuit of birds or wild beasts."[80]

Ormerod says that one of the most interesting figures of Greek legend is Nauplius, whose profession of wrecker, slaver and pirate may be regarded as typical of the early inhabitants of the Mediterranean coast. On shore he is a wrecker, accustomed to lure sailors to their death by means of false flares. At sea he is a slaver and pirate and to him unwanted children and naughty ladies are entrusted to be drowned or otherwise disposed of.

Says Ormerod, "A certain Catreus, king of Crete, gave him his two daughters, Aerope and Clymene, with instructions to sell them into foreign lands. Aerope was sold by Nauplius, but Clymene was retained as his wife. Auge, daughter of Aleos, was similarly handed over for destruction after her liaison with Heracles, and disposed of to a crew of Carian pirates. His name means simply 'sailor' (as the first sailor he is credited with the discovery of the Great Bear), and his conduct probably differed little from that of all early seamen in the Mediterranean. We have already examined the practices of the Mainotes, who were wreckers and pirates in the seventeenth and eighteenth centuries, and of the Tauri in the Black Sea; we hear of other communities who made a live-lihood by such means, where the character of the coast-line was favorable."

§§§

Diffusionism Versus Isolationism

Unfortunately, today's academic enviroment does not allow for the freethinker to speculate a great deal on transoceanic traffic and diffusionism in general. In the university enviroment of to-day there is some danger in speculating about transoceanic con-tact and its ramifications. For the last 70 years or more the label given to those who champion diffusionism and contacts in pre-Columbian times from across the Atlantic and the Pacific is "rac-ist."

Essentially, in these politically correct times, to say that civili-zations in the Americas didn't progress independently from other civilizations in Asia, Africa and Europe is saying that these Ameri-can civilizations were inferior. They had to have "foreign" help from ancient seafarers to build their pyramids, get their purple dye, develop writing, and so on. That American civilizations were somehow incapable of developing such acoutrements of civiliza-tion on their own is a "racist" theory, and therefore should be discounted. Ancient contract and trade via ships in other parts of the world is not deemed a racist concept, but when applied to the Americas, it takes on this tone. It is a twisted bit of logic, but one that pervades the halls of academia to its very foundations.

Anthropologist George Carter had this to say about the racist accusations leveled toward Diffusionists:

If man was voyaging back and forth across the vast Pacific—it spans 1/3 the world—how about man crossing the puny Atlantic? Well, let's be racists. Who? Those dumb Europeans who were primarily still row-ing their galleys around into the 18th century, and who were amongst the last to take up such nautical ad-vancements as stern rudders, and fore and aft rigged sails? Well, fortunately, the Atlantic is so small and the winds and currents flow so strongly and persistently from the mouth of the Mediterranean to the Caribbean that there is no escape. Even a ship wreck will make it in short time. In modern times men have crossed it on a hay stack (Heyerdahl) and a Scandinavian did it on a canvas craft. Even if the Europeans didn't want to, they would arrive in America now and then. The evidence now accumulating is that they wanted to, and did get here, and probably messed up American Indian culture history considerably, though we still can't see, because we haven't tried to see, just when, and where, and how much.

To toss in a few straws to indicate the wind direction: The Spaniards were given European coins when they reached Mexico. One of the Mexican idols had a helmet just like an old helmet worn by one of Cortés soldiers. A pottery head found in Mexican archeology has been identified as made in Rome in the second century A.D., and I have written a little article pointing out that Mexican cylinder seals and many stamps carry alphabetical inscriptions. Well, if all this, one is looking at considerable contact. Did anyone go home? Was this possibly a two-way trade? My answer is: Yes, trade and deliberate voyages, and possibly even colonization.

At Pompeii are portrayals of pineapples. Pineapples are strictly American in origin. So, a pottery head from Pompeii (Rome) and an American pineapple to Pompeii—and the dates match. There is another curious one. The Jews that died at the time of the great revolt against Rome (end of the first century A.D.) in part had fled to the desert. In caves their clothing was preserved so well that even the colors lasted. A study of these dyes showed that one of them was cochineal, and cochineal is produced by an aphid like insect that grows on cactus. Cactus is strictly American. This like so many of these items needs some further work, but that is the probable answer as of this moment. Was there then a trade in dyes? Well the Phoenicians (the name means purple) traded in purple dyes that they made from a shell fish. In America, the natives made the same purple from the related shell fish. This was once fluffed off as: The Spanish taught them. But we have C-14 dates for such dyes as early as 200 B.C. Is this a reasonable trade item? It certainly is. Distant trade must rest on light weight, low bulk, high value materials. Dyes fit this perfectly. What else?

In the bronze age, men began going to the ends of the earth for metals: copper, tin, gold, silver. Southern New England is, or was, rich in copper. In southern New England there is a wealth of evidence of inscriptions, standing stones, huge stones set up on lesser stone, large stone chambers. All of this would fit a bronze age, or megalithic European pattern, and it probably does.

The trend of the data is to show that man was crossing the great oceans far earlier that we have thought. I was chastised twenty years ago for suggesting that this could have begun as early as 2500 or even 3000 B.C. Now we have an African plant being grown in America around 7000 B.C. (*Lagenaria siceraria,* the dipper gourd) and one of my friends is casually mentioning the possibility that men were crossing the Atlantic as early as perhaps 15,000 years ago. THAT is a high date even for me, but give me a month or so to digest it and I may begin to toy cautiously with the possibility.

And that is where we stand. We know with considerable certainty that voyages and meaningful exchanges of plants and animals occurred, and that trade is very probable, and that colonization is quite possible. We have lots to learn and a lot of stubborn people doing their best to block serious inquiry. In too many cases, this leaves the field to what is known as 'the lunatic fringe' and since there are incautious souls out there they often give the professionals the very ammunition they need to smear everyone. If the whole thing weren't so exciting, and so important for the understanding of the origins of civilization and the nature of man, one would be tempted to give it up and just let the Phuddies, to borrow Harold Gladwin's phrase, own the field. But, it is all too important for that.[19]

Carter, in his last sentence, was referring to the book *Men Out of Asia*,[83] written in 1947 by Dr. Harold Gladwin. Gladwin's book is the classic attack on the academic dogma of isolationism and it was in his book that the term "Phuddy Duddy" was coined, referring to smug experts with Ph.D.'s who sit in their high academic chairs pronouncing what is correct as far as history is concerned.

While Gladwin's book made some inroads at the time and the term Phuddy Duddy has been popularized, the dogma of isolationism remains, and, as George Carter points out, it has now taken on the repugnant slant of calling those who favor a diffusion of ideas through various migrations and explorations by sea in antiquity as "racist." This is not a label that any researcher, archaeologist or historian would want to accept, and it therefore makes the entire subject rather volatile and tricky.

We could turn the argument around and say that it is racist to suggest that Chinese, Burmese, Libyans and Jews were too simple minded to have crossed the ocean. History and truth are matters of scientific investigation and evidence and are not dependent on what is politically correct today.

The racist argument is an interesting one, as well, in the sense of pirates and ancient seafarers: the crews of ships of all types were usually a mixed bag of various nationalities. This was advantageous in many ways, as sailors on board a ship who spoke different languages and knew special areas of a sea that the other sailors were unfamiliar with were highly valued. As we shall see later, pirates in particular were an exceptionally democratic and unbigoted group. Where segregation and a caste system were prevalent in the various nations of the world, things were quite different on board a ship. Equality prevaled in regard to color or creed or social status in life—what truly mattered to other sailors was ability, talent and craft. The beginnings of democracy began, incredibly, onboard pirate ships!

And so we see that sophisticated seafaring is many thousands of years old, and piracy is just as ancient. Let us move on to a more recent time, a time of war and religious suppression of science and discovery. It was called the Dark Ages. It was a time peopled by Popes and kings, barons and feudal lords, crusaders and sultans. It was the time of the Knights Templar and their fleet.

The eastern Mediterranean.

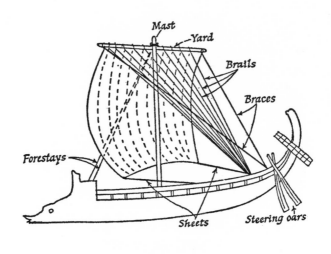

Top: An early Greek ship. Bottom: Arrangement of two banks of oars.

The Seal of the Knights Templar: Two riders on one horse.

Naval battles were so popular in Roman times that the Colliseum was sometimes flooded by diverting a nearby river so that mock naval battles could be staged.

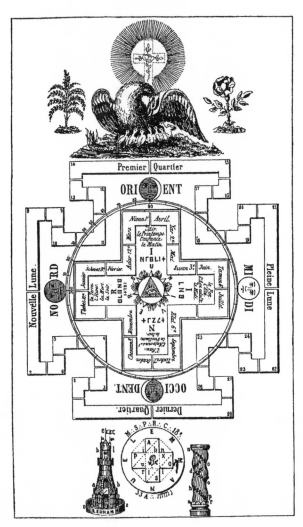

Mystical diagram of Solomon's Temple, as prophesied by Ezekiel and planned in the building scheme of the Knights Templar. The two pillars represent Jachin and Boaz from the original Temple of Solomon. The pillar on the right resembles the Apprentice Pillar at Rosslyn Chapel.

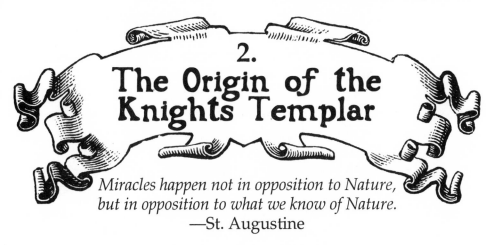

2.
The Origin of the Knights Templar

Miracles happen not in opposition to Nature,
but in opposition to what we know of Nature.
—St. Augustine

Always do right—this will gratify some and astonish the rest.
—Mark Twain (1835-1910)

Since this book is essentially how the modern mythos and iconography of pirates has come from the lost fleet of the Knights Templar, it is important to have a brief history of the Knights Templar themselves.

The Origin of the Knights Templar

The search for the Knights Templar is as fascinating a story as history provides us. The Knights Templar have been associated with all sorts of amazing activities including having the Ark of the Covenant, the Holy Grail, a secret fleet that sailed the oceans, and an awe-inspiring self-confidence and courage that made their enemies fear them.

Despite their fearsome, battle-hardened reputation, the Knights Templar were learned men, dedicated to protecting travellers and pilgrims of all religions, not just Christians. They were great statesmen, politically adept economic traders, and they were apparently allied with the great sailor-fraternity that had created a worldwide trading empire in Phoenician times.

Despite a great deal of negative publicity against the Templars at the time of their suppression, they are still known today as the preservers of knowledge and sacred objects. While the origins of the Knights Templar are said to go back to the building of King Solomon's Temple by Phoenician masons from Tyre, or even to the Great Pyramid and Atlantis, we trace their modern history to the period of the Crusades in the Middle Ages.

In his book *The Mysteries of Chartres Cathedral*[8] the French architect Louis Charpentier claims that the Knights Templar built Chartres as a repository for ancient wisdom, a repository that is equal to Stonehenge, the Temple of Solomon or the Great Pyramid of Egypt. He further claims that special knowledge about the Temple in Jerusalem was gained by the founding group of nine knights during their residence at Solomon's Temple starting in 1118 AD. In that year it is historically recorded that nine "French" knights presented themselves to King Baldwin II of Jerusalem, a Christian king, and explained that they planned to form themselves into a company with a plan for protecting pilgrims from robbers and murderers along the public highways leading into Jerusalem. King Baldwin II had been a prisoner of the Saracens and knew of their disunity. Factions such as the Assassins were also active in Moslem politics.

The knights asked to be housed within a certain wing of the palace, a wing which happened to be adjacent to the Dome of the Rock mosque which had been built on the site of Solomon's Temple. The king granted their request and the order of the Poor Fellow-Soldiers of Christ and the Temple of Solomon (or Knights Templar) was born. Ten years later the knights presented themselves to the Pope, who gave his official approval to the Knights Templar.

Although there had been only nine mysterious knights, a tenth joined them, who was the Count of Champagne, an important French noble. In fact, none of the "poor" knights was apparently poor, nor were they all French. Several came from important

French and Flemish families. Of the ten original knights, four have not been fully identified, although their names are known.

The ranks of the Knights Templar grew rapidly. When a nobleman joined their ranks, he surrendered his castle and property to the Knights, who would use revenues generated from the property to purchase weapons, war-horses, armor and other military supplies. Other noblemen and Kings who were not members often gave them gifts of money and land. King Steven of England contributed his valuable English manor of Cressing in Essex. He also made arrangements for high-ranking members of the Knights to visit nobles of England and Scotland. By 1133, King Alfonso of Aragon and Navarre (northern Spain) who had fought the invading Moors in 29 battles, willed his kingdom to the Templars. However, the Templars were prevented from claiming the kingdom because of the Moorish victory over Spain.

Pope Eugenius decreed that the Knights Templar and only the Knights Templar would wear a special red cross with blunt, wedge-shaped arms (called the cross patee) on the left breasts of their white robes, so that they could be quickly recognized at any time by Christians and by other Templars on the field of battle. The white robes with red crosses became their required dress. The warrior-knights fought bravely in the Middle East, and were highly respected by their Moslem counterparts for their strategy and bravery. Many Templars were, in fact, of Palestinian birth, spoke perfect Arabic, and were familiar with every religious sect, cult and magical doctrine. For instance, the Grand Master Philip of Nablus (1167 AD) was a Syrian.[7]

With the help of the brilliant French abbot Bernard de Clairvaux, the knights, directed by the Count of Champagne, had become well established. The Templars proved a very successful group, distinguishing themselves on the battlefield and amassing great fortune. But what were they really all about? Charpentier likens the mission of the original band of Knights Templar to a commando raid on the ancient Temple of Solomon with the purpose of uncovering its engineering secrets, and possibly recovering lost treasure such as the Ark of the Covenant, which may have been hidden deep in a strange cavern system beneath the temple. In fact, with money that they accumulated, the Templars built the fabulous and mysterious cathedral at Chartres. Later, other cathedrals were built around Europe and the legends of the "Master Stonemasons" became common.[8]

Incorporated into Chartres cathedral are beautiful stained-glass windows, many of the colors difficult or impossible to duplicate today. Hidden within the cathedral are various ancient "cubits" of measure, plus such esoteric devices as the famous Chartres Maze and other visual tools such as sacred geometry, for personal transformation—a sort of personal alchemy of the soul. Included in the imagery was the quest for the Holy Grail. Were the Templars themselves on such a quest? It seems somewhat unlikely that the Knights of the Temple of Solomon were formed to protect pilgrims on the path to Jerusalem, because such an order of Knights already existed. They were the Knights Hospitalers of St. John, later to become the Knights of Malta.

§§§

The Knights Hospitalers of St. John

The Knights Hospitalers of Saint John of Jerusalem (generally known as the Knights of St. John), were founded at Amalfi, Italy, in the 11th century. They went to Jerusalem to protect and minister to the Christian pilgrims, but soon extended their mission to tending to the sick and poor all over the Holy Land.

As the years went by the Knights of St. John became increasingly militant and, generally speaking, fought alongside the more mystical Knights Templar and the Germanic order of the Teutonic Knights of St. Mary.

With the fall of Jerusalem in 1187, the Knights of St. John retreated to Acre. After the fall of Acre, the Knights moved in 1309 first to Cyprus and then to Rhodes. As the main base for the crusaders in their struggle against the Ottoman Empire, Rhodes was a fortress, a prison, and a supply base for the ships and armies on their way to Palestine and Asia Minor.

In 1481 when the Ottoman Sultan Mehmet Fatih failed to clarify the succession question of the newly powerful Ottoman Empire, a battle between his two heirs at Bursa resulted and Çem was defeated by his brother Beyazit. Çem fled to Egypt but was denied asylum by the Mamelukes who controlled that country for the Ottomans.

Çem took the irreversible step of fleeing to Rhodes where he availed himself to the archenemies of the Ottomans, the Knights of St. John. With his brother now in the hands of the crusader army, Beyazit knew he was in trouble and the Ottoman Empire had to respond quickly.

Beyazit shrewdly contacted the Knights of St. John and negotiated a contract to pay 45,000 ducats of gold annually—a huge sum at the time—in return for the imprisonment of his brother on Rhodes and later in the English Tower at the castle in Bodrum, on the Turkish mainland.

The Knights eventually handed their valuable prisoner over to the Vatican, where Çem was made an interesting offer: to lead a crusader army to recapture Istanbul (Constantinople).

To stop this final threat from his wayward brother Beyazit spared no expense, paying to the Vatican 120,000 gold ducats and a number of sacred relics from Jerusalem including the famous "Spear of Destiny." The Spear of Destiny was reportedly the spearhead of the Roman centurion Longinus that was used to pierce the side of Jesus while He was on the cross. Another artifact offered was the "sponge of the last refreshment." This was the sponge used to wet Jesus' lips.

Legend has it that the possessor of the Spear of Destiny has the power to rule the world. Adolf Hitler believed in this and removed the spear from the Vienna museum when the Nazis took over Austria.

With this hefty payment, the Pope abandoned Çem and the plans for him to lead an army against Istanbul. Çem died alone at the Terracina prison in 1495. Rumor had it that he was poisoned. Today Çem is but a footnote in history, a victim of the diplomatic maneuvers that brought the Spear of Destiny to the West.

The Knights stayed on Rhodes for 213 years, transforming the city into a mighty fortress with 12-meter thick walls. They withstood two Muslim offenses in 1444 and 1480, but in 1522 the sultan Suleiman the Magnificent staged a massive attack with 200,000 troops.

A mere 600 Knights with 1,000 mercenaries and 6,000 Rhodians eventually surrendered after a long siege, being guaranteed safe passage out of the citadel. For a brief period the Knights of St. John were without a major base until 1529 when Charles V, grandson of Ferdinand and Isabella of Spain, offered Malta to the them as their permanent base. In that year the soon-to-be-called Knights of Malta began to build fortifications around the Grand Harbor of Valetta, Malta's largest port. In 1565 the Ottoman fleet arrived at Malta and immediately attacked the fortifications.

With 181 ships carrying a complement of over 30,000 men, the fleet bombarded the fortress with over 7,000 rounds of ammunition every day for over a month, and finally took St. Elmo. But the Turkish marines had suffered many casualties and could not take the other heavily-defended forts that were around the bay and the interior of the island. News of reinforcements coming from Sicily caused the Turks to retreat from the island and the Great Siege was over.

The Knights of St. John changed their name to the Knights of Malta and were said to be fanatically loyal to the Vatican, and the Pope apparently used them as his personal crusaders and soldiers. Other orders such as the Knights Templar and the Teutonic Knights were far more independent. The Knights of Malta are sometimes even said to be a secret society for the Vatican.

It was Napoleon who finally ended the rule of the Knights over Malta. The Knights had sent money to Louis XVI, and when the French King was executed, the Knights of St. John on Malta were denied any French revenues by Napoleon.

The Knights of Malta then turned to the Russian Tsar Paul I who offered to found an Orthodox League of the Knights of St. James. This deal with the Russian Tsar particularly enraged Napoleon.

Napoleon sailed to Malta and made anchor just outside the Grand Harbor in June of 1798. When he was refused entry by the knights, he began to bombard the fortress. After two days of shelling, the French landed and gave the knights four days to leave, thus ending their 268-year presence on the island.

The Pope restored the office of the Grand Master in 1879 and the Knights of Malta, although they never returned to Malta. Instead they have offices in Rome and various other cities in Europe

(Prague and Vienna, for example). Even though they have no actual territory, they are still recognized as a separate state by 40 or more countries around the world, similar to the recognition of the Vatican.

§§§

The Treasure of the Knights Templar

It is important not to confuse the Knights Templar with the Knights of Malta, as many readers, and some historians, do. The Knights Templar are quite different from the other crusaders, and were sometimes even said to fight in combat against the others, even in the Holy Land.

Charles Addison, a London lawyer, who wrote about the Templars in his 1842 book *The History of the Knights Templars*[6] mentions in the first few pages how it was commonly believed that Templars were at odds with the Vatican and its military arm, the Knights of St. John. Addisson denies the allegations, but demonstrates that such rumors did exist.

It is not so far-fetched to believe that the Knights Templar secretly opposed the Church since their inception. As we queried earlier, what was their actual mission? If they truly wanted to protect the pilgrims in the Holy land, why didn't they simply join the order of St. John, which was already in existence?

As mentioned above, some people believe that the Knights Templar were the Middle Ages version of a society that had its inception far earlier. It was the purpose of this society to secrete and preserve truly valuable items, documents and knowledge, from the swing and sway of changing temporal powers. If this were the case, it would be logical to assume that the knights were following their own agenda—one that would vary greatly from that of the Roman Catholic Church in that era, which would seem to have been to establish hegemony, collect as much money as possible and put fear into the hearts of the people.

For hundreds of years there have been rumors that the Knights Templar were not only the defenders of the true faith, but were also the guardians of the Holy Grail. The Holy Grail is said to be the most holy of religious artifacts. Many versions of the legend exist, and there are two distinctly different interpretations as to what the Holy Grail comprised. Some say it is the chalice used by Christ at the Last Suppers. Others say that Jesus and Mary Magdalene were married and had offspring. The Holy Grail (or Sangreal) protected by the Templars was the genetic blood lineage of Jesus, carried among his heirs. One chalice version of the Holy Grail says that Saint Joseph of Arimathea brought the cup, which had been used to collect the blood that flowed from Christ's wounds, to England. A Welsh version of the Grail story says he left the holy relic at Glastonbury, from whence it reached King Arthur and the Knights of the Round Table. The Sangrael-Holy Blood stories describe the flight of the holy family, following the crucifixion, to the southern France, where the bloodline continued through the Merovingian kings.This included such famous Merovingian kings as Dagobert and Childeric.

Whatever one believes about the true mission of the Knights Templar, the historical fact of their persecution remains the same, and can be explained as stemming from simple greed.

During the 180 years of Crusades, the Templar wealth grew into a huge fortune. They owned over nine thousand manors and castles across Europe, all of which were tax free. Each property was farmed and produced revenues that were used to support the largest banking system in Europe, which they had established. The Templar wealth and power caused suspicion and jealously among some members of the European nobility. Slanderous rumors were spread of secret rituals and devil worship.

After the defeat of the crusaders in the Eastern Mediterranean, the Templars moved from Jerusalem to the Pyrenees with their headquarters in Paris. Their main naval base was La Rochelle, a French port near the Pyrenees. The Knights Templars operated in many European countries and Mediterranean islands with strong links to Portugal, France, Holland, England and Scotland.

Their main base of operations was southern France and Catalunya, the area of the Cathars and the Merovingian Kings. Barcelona, the capitol of Catalunya, was originally a Phoenician port and this area along the border of Spain and France has long thought of itself as a state, people and culture separate from the rest of Spain. The populace speaks its own language, Catalunian, a language that may have originated with ancient Phoenician.

Outside of Barcelona is the mountain named Montserrat, a site of religious pilgrimage for a long time, probably going back even before the Christian era. It rises 4,054 feet above the coastal plain, and eventually became the site of a celebrated Benedictine monastery. It was at Montserrat that Saint Ignatius of Loyola vowed to dedicate himself to a religious life.

According to the *Grolier Multimedia Encyclopedia* (1997), Montserrat was thought in the Middle Ages to have been the site of the the Holy Grail. Says the encyclopedia, "the monastery contains the Shrine of Our Lady of Montserrat (La Moreneta), a wooden statue said to have been carved by Saint Luke. During the Middle Ages, the monastery was the legendary home of the Holy Grail."[29]

Indeed, there is a certain mystery surrounding Montserrat. The monastery makes an almond liqueur called Aromes del Montserrat which uses for its logo a steep mountain peak with a box and cross at the summit.

When visiting the monastery with a group of tourists, I wondered if the box on top of the mountain was the legendary Ark of the Covenant, a wooden and metal box which held the sacred relics of the Temple in Jerusalem. Was the Ark of the Covenant and, later, the Holy Grail, taken to a secret cavern complex in the mountains of Montserrat? Legend says that a secret Brotherhood of Essenes moved to Catalunya and Montserrat in the early centuries before Christ. This secret brotherhood built a network of subterranean caverns, castles and monasteries in the Pyrenees Mountains along the border of southern France and Spain. Did the Templars move to this area in order to safeguard these relics?

§§§

The Persecution of the Knights Templar

What is certain is that King Philip IV of France was growing envious of the Templar popularity and wealth, and was responsible for many of the slanderous rumors spread about them. Philip IV had once taken refuge from an angry mob in Paris at the Templar Headquarters there. The Templars gave Philip IV refuge from the mob, but it is said that the King saw the magnificence of the Templar treasure and wanted it for himself. His nation was bankrupt, he was a weak king who was unpopular with the people, and he knew that the Templars were more powerful and wealthy than he was.

King Philip IV went to Rome in 1307 to convince Pope Clement V that the Knights Templar were not holy defenders of the Catholic faith, but were seeking to destroy it. On October 13, 1307, King Philip ordered simultaneous raids on all the Templar priories in his country. Within the next few days and weeks, hundreds of knights were captured, including the Grand Master of the Order, Jacques de Molay.

Templars throughout Europe began to be arrested and harassed to extract confessions of heresy. The Templars who were captured were all tortured, sometimes for months and years continually, in order to make them divulge the secret of the Holy Grail and its whereabouts. The Grand Master, Jacques de Molay, was given special attention and suffered years of agony. But it is said that none of the Templars revealed the desired information.

In 1312, Philip pressured Pope Clement V into disbanding the Order throughout Europe even though it was already effectively crushed in France itself. A papal bull suppressing the Order was finally issued.

By now, when the King's men went to the Templar castles they found many of them abandoned. Those who were arrested were tried and found guilty of sins against God. In 1314, Jacques de Molay who had suffered in Philip's dungeons for seven years— he'd been blinded by red-hot irons thrust into his eyes, his genitals had been boiled in oil and then pulled off with cords, most of his bones had been broken or dislocated on the rack—faced his last torment: he was roasted alive over a slow fire by order of king and pope. This barbarous destruction of a human being was never to be forgotten, and today de Molay's name lives on, 700 years later, as the title of a Masonic youth group.

A contemporary English poem asked the question that many ask today:

Where did all the Templars and their great wealth go?
The brethren, the Masters of the Temple,
Who were well-stocked and ample
With gold and silver and riches,
Where are they?
How have they done?
They had such power once that none
Dared take from them, none was so bold;
Forever they bought and never sold...

This question has plagued historians and treasure hunters for centuries.

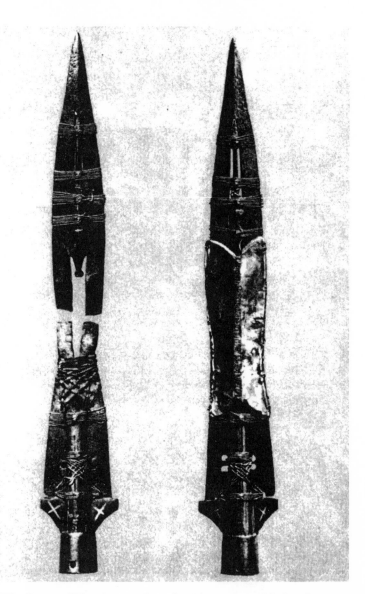

The Spear of Destiny, with and without its gold sheaf.

Secret writing alledged to belong to a surviving branch of the Templars. From Arkon Daraul's book *Secret Societies*.

Knights attacking a castle in the Holy Land.

Crusader ships at the port of Tyre in the Holy Land.

A crusader in the streets of Jerusalem.

Crusader ships and dhows at the port of Acco in the Holy Land.

A wounded knight in the Holy Land.

Burning of the Templars, from the *Chroniques de France ou de Saint Denis.*

The Holy Sepulchre

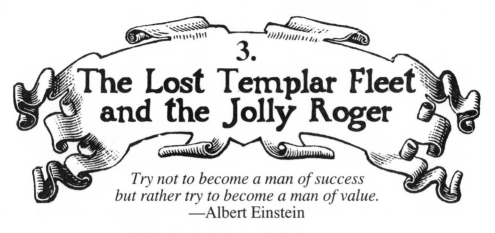

3.
The Lost Templar Fleet and the Jolly Roger

*Try not to become a man of success
but rather try to become a man of value.*
—Albert Einstein

*In three words I can sum up everything
I've learned about life. It goes on.*
—Robert Frost

The Knights Templar, now officially disbanded, were dispersed all over Europe and were hunted men. But, these refugee knights still commanded immense respect from fighting men all over Europe. These Grail Knights were welcomed in many places and given sanctuary as heroes. Some joined the Teutonic Knights and fought against Mongol and Tartar incursions in Eastern Europe; some went to Hungary and fought against Turkish expansion; some went to Scotland, others went to Portugal.

The Lost Fleet of the Knights Templar

The mysterious Knights Templar had an extensive sea network and may even have inherited some of the maps and other secrets of the Phoenicians. The Templar fleet is discussed in Michael

Baigent and Richard Leigh's book *The Temple and the Lodge*.[3] They point out that the Templars had a huge fleet at their disposal, a fleet that was stationed out of ports in Mediterranean France and Italy as well as ports in northern France, Flanders and Portugal:

"On the whole, the Templar fleet was geared towards operation in the Mediterranean—keeping the Holy Land supplied with men and equipment, and importing commodities from the Middle East into Europe. At the same time, however, the fleet did operate in the Atlantic. Extensive trade was conducted with the British Isles and, very probably, with the Baltic Hanseatic League. Thus Templar preceptories in Europe, especially in England and Ireland, were generally located on the coast or on navigable rivers. The primary Atlantic port for the Templars was La Rochelle, which had good communication with Mediterranean ports. Cloth, for example, could be brought from Britain on Templar ships to La Rochelle, transported overland to a Mediterranean port such as Collioure, then loaded aboard Templar ships again and carried to the Holy Land. By this means, it was possible to avoid the always risky passage through the Straits of Gibraltar, usually controlled by the Saracens."[3]

When King Philippe ordered the dawn raids on October 13, 1307, the Templar fleet based at La Rochelle somehow got advance warning. The entire fleet set sail, escaped Philippe's net, and has never been heard from since. This date, October 13, 1307 was a Friday, and this became the origin of popular belief that Friday the 13th was unlucky, as it had been the official day of the suppression of the Templars.

The disappearance of the Templar fleet has become one of the great mysteries of history, what happened to this considerable fleet of ships? Was it scattered amongst the seven seas, or did it regroup at some secret assemblage?

Baigent and Leigh in *The Temple and the Lodge*[3] claim that the Templar fleet escaped en masse from the various ports in the Mediterranean and northern Europe and left for a mysterious destination where they could find political asylum and safety. This destination was Scotland.

The Mediterranean fleet had to sail through the dangerous Straits of Gibraltar and then probably stopped at various Portuguese ports that were sympathetic to the Templars such as Almourol castle. Portugal was one of the few places where they could find some asylum. Portugal was a country which, unlike Spain, was largely sympathetic to the Knights Templar.

Typical of Templar strongholds and maritime fortifications is Almourol Castle near Abrantes in central Portugal. The fortress stands with towers and crenelations on a small rocky island covered in greenery in the Tagus River. In this spectacular setting, the castle has given rise to much romantic lore of knights, seafarers, and pirates. The Portuguese author Francisco de Morais' long prose entitled "Paleirim de Inglaterra" (Palmeirim of England) chronicles duels and swordfights at the foot of the ramparts, while the local giant is named Almourol.

Almourol Castle was constructed by Gualdim Pais, who was a Master of the Order of Templars in 1171 on the foundations of an earlier Roman (and possibly Phoenician) castle. Gualdim Pais was one of many Portuguese nobles who were Master Templars, great navigators, and the possessors of mighty castles and large fleets. The role of Portugal cannot be underestimated or overlooked when researching the lost Templar fleet and the ultimate regrouping of the Templars as Masons, pirates and the Portuguese Knights of Christ.

In Portugal the Knights Templar retained something of a cohesive organization, and merely changed their name to "The Order of the Knights of Christ." Here they found royal support to which the Church could only turn a blind eye. First, King Alfonso IV of Portugal became the Grand Master of the "new" Knights of Christ. Later, Prince Henry the Navigator was also Grand Master of the Templars.[10]

The curious history of Portugal, and later Brazil, is tied into the Templars and their lost fleet. Even the name Portugal is a curious one. It has been suggested that it is a Templar name "Port-O-Gral," meaning "Port of the Grail."

Portugal has always had close ties with England, and one wonders if this does not have something to do with the Knights

Templar and their fleet in Portugal. And how was small Portugal to remain independent on the Iberian peninsula? Was it because of its strong connection to the Templars and ancient seafarers? Later, its transatlantic colony Brazil was set up in much the same way as the United States was set up, as a union of independent states. And it was set up by a series of Masonic lodges just as in North America.

§§§

The Templar's Last Stand

Baigent and Leigh go on to say that the Templar Fleet sailed from Portugal up the west coast of Ireland to the safe ports in Donegal and Ulster, where Templar properties were located and arms smuggling to Argyll was common.

The Templar fleet then landed in Argyll by sailing to the south of the islands of Islay and Jura into the Sound of Jura where the Templars unloaded men and cargo at the Scottish Templar strongholds of Kilmory, Castle Sweet and Kilmartin.

Robert the Bruce controlled portions of Scotland, but not all of it. Significant portions of the northern and southern Highlands were controlled by clans that were allied with England. Robert the Bruce had been excommunicated by the Pope in 1306, one year before the persecution of the Templars began. Essentially, the papal decree that outlawed the Knights Templar was not applicable in Scotland, or at least in the parts that Robert the Bruce controlled.

The turning of the tide for Robert the Bruce, Scotland and (maybe) the Knights Templar was the famous Battle of Bannockburn. Although the actual site of the battle is not known, it is known to have taken place within two and a half miles of Stirling Castle, which is just south of Edinburgh.

On June 24 of 1314, Robert the Bruce of Scotland with 6,000 Scots miraculously defeated 20,000 English soldiers. But exactly what took place has never really been recorded. It is believed by some that he did it with the help of a special force of Knights Templar.

Say Baigent and Leigh in their book: "Most historians concur that the Scottish army was actually made up almost entirely of foot soldiers armed with pikes, spears and axes. They also concur that only mounted men in the Scottish ranks carried swords, and that Bruce had few such men ..."[3]

Suddenly in the midst of combat, with the English forces engaged in a three-to-one battle against the Scottish soldiers, there was a charge from the rear of the Scottish camp.

A fresh force with banners flying rode forth to do battle with the English. The English ranks took one look at the new force and in sheer terror of the new combatants, they literally fled the field. Say Baigent and Leigh, "... after a day of combat which had left both English and Scottish armies exhausted... Panic swept the English ranks. King Edward, together with 500 of his knights, abruptly fled the field. Demoralized, the English foot-soldiers promptly followed suit, and the withdrawal deteriorated quickly into a full-scale rout, the entire English army abandoning their supplies, their baggage, their money, their gold and silver plate, their arms, armour and equipment. But while the chronicles speak of dreadful slaughter, the recorded English losses do not in fact appear to have been very great. Only one earl is reported killed, only 38 barons and knights. The English collapse appears to have been caused not by the ferocity of the Scottish assault, which they were managing to withstand, but simply by fear."[3, 13]

In fact, what appears to have happened was a charge in full regalia by the remaining forces of Knights Templar against the English host. These crusade veterans were like the Green Berets or Special Forces of the Middle Ages. All combatants suddenly stopped to witness the charging army of Knights Templar, white banners with red cross insignias flying high above these mounted Grail Knights. This sight evidently frightened the English forces so much that even though they still had a superior force, they fled, rather than fight.

The probable strategy behind the Templars' charge into battle (inspired by the Muslim Assassins—see Ch. 4) would have been to ride through the thick of the battle and attempt to reach King Edward and his personal guards. Once engaged with the commanding officers of the English foe, these seasoned war veterans would have easily defeated King Edward's knights and possibly killed the king himself. As noted, King Edward and his special knights immediately fled upon witnessing the Templar charge.

And thus Scotland stayed an independent kingdom, free from the domination of Rome and the Pope. It was thus a safe haven for the missing Templar fleet and for the knights themselves.

But what became of this missing fleet? Was it grounded in Scotland? Did it sail across the Atlantic a hundred years before Columbus? Did it become one of the great fleets of the Portuguese and Scottish kings? Did it become a fleet of pirates that attacked ships loyal to the Pope and the Vatican? Perhaps all of the above!

§§§

Pirates—Guardians of the Sangrael

According to Michael Bradley, author of *Holy Grail Across the Atlantic,*[9] the destruction and dispersal of the Knights Temlar caused several repercussions in Europe, the first of which was an upsurge in "piracy."

Bradley believes that the Templars were guardians of the Holy Grail, in the sense that it comprised the "Sangrael" or "Holy Blood." The story goes that descendants of Jesus living in southern France were evacuated from Montségur during a terrible siege in March of 1244. They probably hid in numerous secret Pyrenees caverns for months or even years, as some troubadour poetry refers to.

But for the long-term safety of the Templars and the Holy Blood, the Order and the Grail had to be taken out of France eventually, and, ultimately, even out of Europe. Bradley says that the Garonne River was the obvious route of eventual escape, as it reaches from deep in the Pyrenees across southern France to the Garronne Estuary on the west coast. La Rochelle, home of the Templar fleet, is situated on the Estuary. A sanctuary at the town and fortress of Angouleme along the way may have been used as a haven for two or three generations. So long as the Templars remained a cohesive and independent Order, there was hope that the lineage could successfully hide secretly in Europe and begin the process of recouping its fortunes.

Bradley believes that at least some of the Templar vessels which disappeared in 1307 carried the Holy Grail. This treasure of the Templars may have not only family members of the former Merovingian Royal Family, but also a large assortment of physical treasures, including a fortune in gold coins, crosses and jewelry, plus possibly even the famous Cup of the Last Supper (Grail Chalice) or incredibly, the Ark of the Covenant. Both were rumored to be in their possession, though most historians believe it to be mere fantasy, as they regard most of the Holy Grail material. The famous Spear of Destiny was in the hands of the Knights of Malta, who obtained it from the Ottomans (see chapter 2).

Bradley agrees with Baigent and Leigh that the Templar fleet sailed north to the Irish Sea to their final destination: the western fiords of Scotland, avoiding the Irish coast because of pro-Vatican forces in control of those ports. They landed in various areas north of present day Glasgow, in the environment of Oban. Here Templar graves were marked with the familiar symbols of the skull and crossbones, now known as the Jolly Roger. The renegade Templar fleet was now under control of the St. Clair (Sinclair) family of Rosslyn and the naval war between the Templars and the Vatican was to begin in earnest.

The fleet was to be split up into two different parts, one part would anchor in the Orkney Islands and prepare for Sinclair's ambitious journeys to North America. The other part of the fleet was to fly the skull and crossbones and sail the Atlantic and Mediterranean as a naval force that would attack those ships that were seen as being associated with the Vatican and the kings that supported the Roman Catholic Church.

Says Bradley about the Templar ships that were to become the early pirates, "Other vessels, however, seem to have been used as the Blood's navy, to strike back at the hated Roman Church and the monarchs and countries loyal to it. An upsurge in European piracy begins from this time and the pattern of it suggests that many pirates were not mere freebooters who would attack anyone, but very curious 'pirates' who confined their attentions to Vatican and loyal Catholic shipping."[10]

It would be later that such British seamen as Captain Drake (whose name means "dragon") would elevate "piracy" to big business, but he preyed only upon Spanish-Catholic ships once the Inquisition had been installed in Spain. When the Spanish Inquisition was installed in the New World after 1492, the Templar "pirates" spread their attacks to the Caribbean and eventually even to the Pacific ports in Peru and Mexico, all in the name of a naval war being waged for over 200 years.

This was the mystique of the early pirates: they were essentially a secret navy that was not supposed to exist. This navy had even been to the New World nearly a century before the Spanish. The flag of this navy was to become the symbol of Masons and pirates alike: the skull and crossbones—known worldwide as the Jolly Roger.

§§§

The Kingdom of the Jolly Roger

The first known historical case of a fleet of ships flying the black flag of the skull and crossbones was King Roger of Palermo, also called Roger of Sicilly and "Jolly Roger." King Roger was a Templar from Normandy who conquered Sicily during the time of the Kingdom of Jerusalem.

Because the Templar fleet could not fly their original flag of a red cross on a white background, they began to fly a new flag, essentially the war flag of an almost vanquished navy. While some of the Templar fleet was absorbed into the Portuguese navy and flew the Portuguese flag, the other ships, usually commanded by French or English captains, flew a black flag with a skull and crossbones on it.

The Jolly Roger is named after Roger II of Sicily (1095-1154). Roger had reputedly been associated with the Templars during the crusades, and conquered Apulia and Salerno in 1127 AD, despite opposition from Pope Innocent II. His court was full of dancers, music and entertainers, and he was known as "Jolly Roger." His fight with the Pope was well known, especially to seafarers and traders.

The Kingdom of Jerusalem was set up by the Knights Templar as a Christian kingdom, controlled by the Templars and their nominated king. The first King of Jerusalem was Baldwin I, the brother of Godfrey of Bouillon, whom he accompanied on the first crusade. Godfrey of Bouillon (1058-1100) was a duke of Lower Lorraine and the leader of the first Crusade which later spawned the Knights of St. John or Knights Hospitalers (later to become the Knights of Malta, see chapter two). Godfrey of Bouillon was elected king of Jerusalem, which he helped capture, but took the title "protector of the Holy Sepulchre."

Baldwin I took the title of King of Jerusalem in the year 1100, the year in which all the chief ports in Palestine had been gained by the Templars and the Knights of St. John.

His cousin and successor, Baldwin II (Baldwin du Bourg) fought the Turks in northern Syria and made Tyre and Antioch part of the Kingdom of Jerusalem. He probably also made pacts with the Assassins.

Baldwin II died in the year 1131 and was succeeded by Baldwin III (son of Fulk of Anjou) who reigned from1130 to 1162. The Crusader Latin states in the eastern Mediterranean were beginning to wane and Edessa fell to the Muslims in 1144 and the second crusade ended in failure with the Sultan Nureddin siezing northern Syria in 1154.

The Kingdom of Jerusalem was in full decline when Baldwin IV (the Leper) reigned 1174 to 1183. He spent his whole reign defending the kingdom against Saladin and his evergrowing army and navy. Disabled by leprosy, he made Guy of Lusignan his lieutenant, but withdrew the commission and had his five-year-old nephew crowned in 1183 as Balwin V who reigned until the collapse of the kingdom in 1186.

During this time, two Norman brothers, Roger and Robert Guiscard, carved a small Templar kingdom out of Sicily, Apulia and Calabria in what is today southernmost Italy. Apulia and Calabria were former Byzantine states while Sicily had been controlled by north African Arabs from1057. Wars in this area of southern Italy and Sicily continued nearly unabated until Sicily was freed on the Muslims

in 1091.

Robert Guiscard (c.1015-1085) is today known as the Norman conqueror of southern Italy. He was the son of Tancred de Hauteville, a Norman nobleman. His brothers William Iron Arm, Drogo, and Humphrey Guiscard had preceded him to Italy and had conquered Apulia by the year 1042. Robert succeeded as count of Apulia in 1057, and in 1059 was invested by the pope with this duchy and southern Italian lands the Normans had acquired (or soon were to) from the Byzantines and Arabs.

Robert proclaimed his brother, Roger, count of Sicily and Calbria in 1072. In 1081, Robert began his expedition against what was left of the Byzantine Empire. He took Corfu and defeated Alexius I but returned in 1083 to aid Pope Gregory VII against Emperor Henry IV. He briefly held and sacked Rome in 1084. He was expelled that same year by Henry's forces and resumed his campaigns in the eastern Mediterranean, dying of fever in 1085 on the island of Cephalonia. His possessions in southern Italy passed to his son and grandson until they were annexed into the court of "Jolly" Roger, or King Roger II of Sicily.

Count Roger I died in 1101 leaving Sicily to his son and successor, Roger II (c.1095-1154). This Roger was to become the "Jolly Roger" of history, and the mythological patron of the lost Templar fleet and the pirates to follow.

It was sometime during this period, when the Norman brothers' fleet of ships ranged from northern France to Sicily and the Aegean Sea, that they first flew the skull and crossbones on ships. Certainly King Roger II employed the skull and crossbones on his ships.

Roger II continued the Norman conquests in Sicily and southern Italy, and conquered Apulia and Salerno (1127) despite opposition from Pope Innocent II. He was crowned king of Sicily by Antipope Anacletus II in 1130. Innocent eventually yielded and invest Roger with the lands he already possessed. Roger II established a strong central administration and was famous for his brilliant court at Palermo as a center for arts, letters, and sciences. In may ways, the beginnning of the the Rennaisance occurred at "jolly" Roger's court in Palermo.

Roger II also established a navigational school on Sicily, one isolated from the Vatican by both the sea and his fleet, which invited Jewish and Islamic geographers as consultants. The Arabic geographer, Ibn Idrisi, was attracted to Roger's court and produced a "celestial disc" and a "terrestrial disc," both in silver, which represented respectively all the astronomical and geographic knowledge of the day. Idrisi and Roger also produced the important navigational treatise called the *Al Rojari*. Roger was fond of Islamic and ancient Hebrew love-poetry, with a taste for beautiful women to match the allusions in it.

Roger, through the Arab geographers in his court, also began to get hold of important maps that the Muslims had aquired from libraries such as the one at Alexandria in Egypt. More on that in chapter 5.

§§§

The Rise of Piracy After the Suppression of the Templars

Michael Bradley maintains that the Templar dispersion had several immediate repercussions in Europe of which an upsurge of "piracy" in the 14th and 15th centuries was the first.

A second repercussion was that wherever the Templars dispersed with some cohesion, voyages of Atlantic exploration immediately followed. In Portugal, the Templars didn't go underground at all, but merely changed their name to the "Knights of Christ" and King Alfonso IV immediately became this "new" Order's Grand Master. Is it mere coincidence that King Alfonso IV also immediately began the Portuguese policy of sending ships out into the Atlantic?

Prince Henry the Navigator continued his father's tradition of Atlantic exploration. He was also a Grand Master of the Knights of Christ, the new Templar Order in Portugal.

Says the German historian Paul Herrmann in his *Conquest By Man*: "King Alfonso IV (1325-57) seems to have initiated long voyages to the west, probably the Canaries, as early as the first quarter of the 14th century. This tradition was taken up and continued by Prince Henry the Navigator (1394-1460) with the aim of finding a seaway to India round the southern tip of Africa."

It was also noted that Templars fled to Scotland and were able to maintain some cohesion there under the protection of the famous Sinclairs. There is a Templar cemetery on the Saint-Clair domain at Rosslyn, which is a tourist attraction today. In Scotland, the Templars did not apparently try to organize themselves into a new Order of knighthood but elected instead to spread their secret doctrine more widely by creating Freemasonry, again under the leadership of the powerful Saint-Clairs. Is it sheer coincidence that voyages to the west set out from Scotland within a generation after Templars had established themselves there?

After the persecution of the Templars and the disappearance of their fleet, pirates were plying the Atlantic and the Mediterranean under their trademark: the Jolly Roger. They only attacked Spanish, French, and Italian ships, or others allied to Rome. Portuguese ships were largely safe—because it was known that Portugal was a safe haven for Templars?

Authors such as Baigent, Lincoln and Bradley maintain that the "skull and crossbones" is carved on many Templar and Freemason gravestones and is nothing more or less than the old Templar "cross pattee" rendered in human skeletal material, with the knobs of the leg-bones being the pattees of the Templar cross. The message of the skull and crossbones is abundantly clear says Bradley: a "neo-Templar" vow to oppose the Roman Church to the death, and thus the symbolism of human bones on both the flag and the Templar and Freemason gravestones.

And so the Templar fleet was divided into four divisions: as the fleet of Jolly Roger of Sicily; as the Portuguese Navy under the Knights of Christ; as pirates flying a black flag with a skull and crossbones; and as a Scottish-Norwegian fleet that would sail from the Orkneys to the New Scotland in what is today Canada.

The secret naval war with the Vatican had begun. A war that was to encompass transatlantic empires, vaste treasure troves and pirate attacks that involved entire fleets of ships. It was a battle of two flags. Which flag would fly over the seven seas?

On top of all this, the Templars had ancient maps that showed some of the lands across the Atlantic. But where did they get these maps which inspired them to make the long voyages into the unknown? This question leads us to the maps of the ancient seakings. But first, let us look at a rival of the Templars: the Assassins.

De Molai in the Torture Room.

A Templar in a French dungeon.

The port of La Rochelle in southwestern France.

The giant that guards Almourol Castle in Portugal.

Jacques de Molay is arrested and tortured.

The Sinclair Knights Templar tombs in Danbury Church, Scotland.

Mary Magdallen holds up an egg as a symbol of Christ. It is her children that are part of the Holy Grail.

Pirates flew the Jolly Roger.

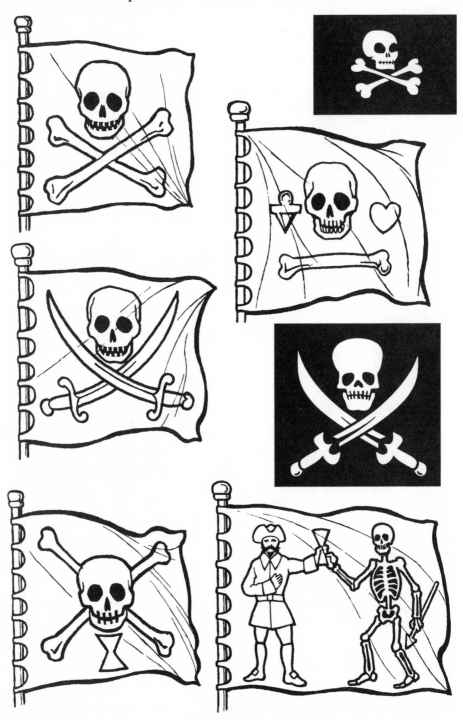

Pirates flew the Jolly Roger, flags which had navigation and Masonic symbols on them.

Left: The Sinclair family arms. Right: Templar grave in Scotland marked with a skull and crossbones.

Templar graves in Scotland.

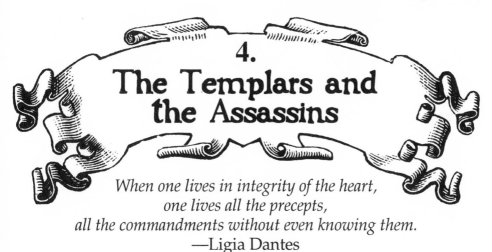

4.
The Templars and the Assassins

When one lives in integrity of the heart,
one lives all the precepts,
all the commandments without even knowing them.
—Ligia Dantes

The fastest way to succeed is to look as if you're playing
by other people's rules, while quietly playing by your own.
—Michael Korda

In some ways, in order to understand how the lost Templar fleet ultimately became the pirates of yesteryear, we must look at the odd association between the Templars and a secretive sect of Islam known as the Assassins (an Arabic word meaning "hashish eaters").

The Assassins, like the Templars, have had a strong influence on history and western civilization. It was the Assassins who developed and refined what would be called today a "decapitation"strike. Rather than sending thousands of soldiers to their death to create a new political order or remove an unwanted political figure from power, they sent one (or several) assassins out to do the work. With the leader of an army or district murdered, the entire army would often retreat and the battle would be won without a great loss of life.

The Templars were to learn first-hand the daring strategy of the Assassins during the crusades, and apparently they absorbed some of their techniques into their own methodology. The Templars demonstrated a "decapitation" move in the Battle of Bannockburn when they rode out toward the end of the battle and rode straight for the English royalty (who were watching the battle from the top of a nearby hill) intent on taking down King Edward IV. The English, including the king, fled the battlefield rather than be "assassinated" by the Knights Templar contingent fighting for Scotland.

§§§

The Origins of the Assassins

The infamous Assassin cult was an offshoot of the Shia branch of Islam, an Ismaili form of Islam known as Nizari Islam. It was founded in the mid-11th century by an Arab-Persian named Hasan-i-Sabah ("Hasan, son of Sabah").

Hasan-i-Sabah was born around 1055 and died in the year 1124. He was the son of a Shia Muslim named Ali Sabah, who lived in Qum, Persia, and claimed that his ancestors were Arabs from Kufa in Iraq. Since the governor of the Province was a Sunni Moslem, Ali Sabah spared no effort to assume the guise of a Sunni, since Shia Moslems were persecuted. Eventually, Ali Sabah retired into a monastic retreat, and sent his son Hasan-i-Sabah to an orthodox Sunni Moslem school. This school was no ordinary one. It was the circle of disciples presided over by the redoubtable Imam Muwafiq. It was said that every individual who enrolled under him eventually rose to great power.

According to Arkon Daraul in his famous book *A History of Secret Societies*,[7] while at the school, Hasan-i-Sabah met Omar Khayyám, the tentmaker-poet and astronomer, later to be the poet laureate of Persia. Another of his schoolmates was Nizam-ul-Mulk, who rose from peasanthood to become prime minister. These three made a pact, according to Nizam's autobiography, whereby whichever rose to high office first would help the others. Nizam was the first to rise to power, first becoming a courtier, and later Vizier to Alp-Arslan, the Turkish sultan of Persia.

Nizam helped Omar, in accordance with his vow, and secured him a pension, which gave him a life of ease and indulgence in his beloved Nishapur, where many of his *Rubá'iyát* poems were written. Meanwhile Hasan-i-Sabah remained in obscurity, wandering through the Middle East, waiting for his chance to attain the power of which he had dreamed. After Arslan the Lion died, he was succeeded by Malik Shah and Nizam was to become his Vizier. It was at this time that Hasan-i-Sabah presented himself to Nizam, demanding to be given a place at court. Delighted to fulfill his childhood vow, the Vizier obtained for him a favored place, and relates what transpired thus in his autobiography: "I had him made a minister by my strong and extravagant recommendations. Like his father, however, he proved to be a fraud, hypocrite and a self seeking villain. He was so clever at dissimulation, that [he seemed] to be pious when he was not, and before long, he had somehow completely captured the mind of the Shah."[7]

Malik Shah was young and inexperienced, and Hasan-i-Sabah was trained in the Shiite art of winning people over by apparent honesty. Nizam was still the most important man in the realm, with an impressive record of honest dealing and achievements, and he was becoming suspicious of Hasan-i-Sabah and his motives.

In the year 1078, Malik Shah asked for a complete accounting of the revenue and expenditure of the empire, and Nizam told him that this would take over a year. Hasan-i-Sabah, on the other hand, claimed that the whole work could be done in forty days, and offered to prove it. The task was assigned to him and the accounts were prepared in the specified time but something went wrong at this point. The balance of historical opinion holds that Nizam struck back at the last moment, saying "By Allah, this man will destroy us all unless he is rendered harmless, though I cannot kill my playmate."[7]

Whatever the truth may be, it seems that Nizam managed to have such disparities introduced into the final calligraphic version of the accounts that when Hasan-i-Sabah started to read them they appeared so absurd that the Shah, in fury, ordered him to be exiled. As he had claimed to have written the accounts in his own hand, Hasan-i-Sabah could not justify their incredible deficiencies.

Hasan-i-Sabah had friends in Isfahan, where he immediately fled. There survives a record of what he said there, which sheds interesting light upon what was in his mind. One of his friends, Abu-al-Fazal, notes that Hasan-i-Sabah, after reciting the bitter tale of his downfall, shouted these words in a state of uncontrollable rage: "If I had two, just two, devotees who would stand by me, then I would cause the downfall of that Turk and that peasant."

Fazal concluded that Hasan-i-Sabah had gone mad with visions of revenge. Hasan-i-Sabah insisted that he was working on a strategy to regain his power and set off for Egypt, there to incubate his plans.

Fazal was to become one of the devotees of the Assassin chief, and Hasan-i-Sabah, two decades later, reminded him of that day in Isfahan: "Here I am at Alamut, Master of all I survey: and more. The Sultan and the peasant Vizier are dead. Have I not kept my vow? Was I the madman you thought me to be? I found my two devotees, who were necessary to my plans."[7]

Hasan-i-Sabah had been brought up in the secret doctrines of Ismailism, and recognized the possibilities of power inherent in such a system. He knew that in Cairo there was a powerful nucleus of the Ismaili religious society. And, according to his devotee Fazal, he already had a plan whereby he could turn these people into disciplined, devoted fanatics, willing to die for a leader.

Fazal claimed that Hasan-i-Sabah had decided that it was not enough to promise paradise, fulfillment, and eternal joy to people. He would actually show it to them; give them a taste of heaven in the form of an artificial paradise, where beautiful women played music and fountains gushed sweet-scented waters, where every sensual wish was granted amid beautiful flowers and gilded pavilions. At his fortress of Alamut in northwest Persia, this is what he eventually did. But first he was to make his historic journey to Cairo.

Details of the journey, and the mentality of Hasan-i-Sabah are given in what is supposed to be an autobiographical account of his early days. He was, he says, reared in the belief of the divine right of the Imams, by his father. He early met an Ismaili missionary (Emir Dhareb) with whom he argued strenuously against the Emir's particular form of the creed. Then, some time later, he went through a bout of severe illness, in which he feared to die, and began to think that the Ismaili doctrine might really be the road to redemption and paradise. If he died unconverted, he might be damned. Thus it was that as soon as he recovered he sought out another Ismaili Imam, Abu Najam, and then others. Eventually he arrived in Egypt, to study the creed at its headquarters, the Abode of Learning.

He was received with honor by the Caliph, due to his former position at the court of Malik Shah. The Egyptian Ismailis were excited to receive such a high-profile convert, and the high officials of the court gave the situation a good deal of public play. But his involvement with the Ismailis seemed in the end to help Hasan-i-Sabah more than it did them. As usual, he became involved in political intrigue and was arrested, then confined in a fortress. Legend says that no sooner had he entered the prison than a minaret collapsed, and in some unexplained way this was interpreted as an omen that Hasan-i-Sabah was in reality a divinely protected person.

The Caliph, hurriedly making Hasan-i-Sabah a number of valuable gifts, had him put aboard a ship sailing for northwest Africa. A tremendous storm blew up, terrifying the captain, crew and passengers alike. Prayers were held, and Hasan-i-Sabah was asked to join. He refused. "The storm is my doing; how can I pray that it abate?" he asked. "I have indicated the displeasure of the Almighty. If we sink, I shall not die, for I am immortal. If you want to be saved, believe in me, and I shall subdue the winds."

At first the offer was not accepted. Presently, however, when the ship seemed on the point of capsizing, the desperate passengers came to him and swore eternal allegiance. Hasan-i-Sabah was still calm, and continued so until the storm abated. The ship was then driven onto the coast of Syria, where Hasan-i-Sabah disembarked, together with two of the merchant passengers, who became his first real disciples.

Hasan-i-Sabah was not yet ready for the fulfillment of his destiny as he saw it. For the time being, he was traveling under the guise of a missionary of the Caliph in Cairo. From Aleppo he went to Baghdad, seeking a headquarters where he would be safe from interference and where he could become powerful enough to expand. Back to Persia the road led him, traveling through the country, making converts to his ideas, which were still apparently strongly based upon the secret doctrines of the Egyptian Ismailis. Everywhere he created a really devoted disciple ("fidayi") he bade him stay and try to enlarge the circle of his followers. These circles became hatching-grounds for the production of 'self-sacrificers,' the most promising initiates who were drawn from the ranks of the ordinary converts. Thus it was that miniature training centers, modeled upon the Abode of Learning, were brought into being within a very few months of his return to his homeland.

During his travels, a trusted lieutenant—one Hussein Kahini—reported that the fortress of Alamut was situated in an ideal place for proselytism. Most of the ordinary people of that place, in fact, had been persuaded into the Ismaili way of thinking. The only obstacle was the Governor—Ali Mahdi—who looked upon the Caliph of Baghdad as his spiritual and temporal lord. The first converts were expelled from the country. Before many months, however, there were so many Ismailis among the populace that the Governor was compelled to allow them to return.

§§§

Alamut: The Fortress of the Assassins

Alamut was a nearly impregnable fortress situated in a hidden valley, and Hasan-i-Sabah agreed it was the perfect site for his artificial paradise. The prospective owner of Alamut decided to try a trick. He offered the local Governor three thousand pieces of gold for "the amount of land which could be encompassed by the hide of an ox." When Mahdi agreed to such a sale, Hasan-i-Sabah produced a skin, cut into the thinnest possible thongs, and joined them together to form a string which encompassed the castle of Alamut. Although the Governor refused to honor any such bargain, Hasan-i-Sabah produced an order from a very highly placed official of the Seljuk rulers, ordering that the fortress be handed over to Hasan-i-Sabah for three thousand gold pieces. It turned out that this official was himself a secret follower of the "Sheikh of the Mountain," as Hasan was to become known,.

The year was A.D. 1090. Hasan-i-Sabah was now ready for the next part of his plan. He attacked and routed the troops of the Emir who had been placed in the governorship of the Province, and welded the people of the surrounding districts into a firm band of diligent and trustworthy workers and soldiers, answerable to him alone.

Within two years the Vizier Nizam-ul-Mulk had been stabbed in the heart by an assassin sent by Hasan-i-Sabah. Hasan-i-Sabah's desire for revenge upon his class-fellow made him the very first target of his reign of terror. The Emperor Malik Shah, who dared to send troops against him, died a mysterious death, and poisoning was gravely suspected.

So by 1092, Hasan-i-Sabah was the powerful head of a strong and growing Shiite sect, known officially as the Nizari-Ismailis. But they were becoming better known as the Assassins.

With the king's death, the whole realm was split up into warring factions. For long the Nizari Assassins alone retained their cohesion. In under a decade they had made themselves masters of all of Persian Iraq, and of many forts throughout the empire. This they accomplished through secretive forays, direct attacks, use of the poisoned dagger, and any other means which seemed expedient. The orthodox religious leaders pronounced one interdict after another against their doctrines; all to no avail.

By now the entire loyalty of the Ismailis under him had been transferred from the Caliph to the personality of the Sheikh of the Mountain, who became the terror of every prince in that part of Asia, the crusader chiefs included. "Despite and despising fatigues, dangers and tortures the Assassins joyfully gave their lives whenever it pleased the great master, who required them either to protect himself or to carry out his mandates of death. The victim having been pointed out, the faithful, clothed in a white tunic with a red sash, the colors of innocence and blood, went on their mission without being deterred by distance or danger. Having found the person they sought, they awaited the favorable moment for slaying him, and their daggers seldom missed their aim."[7]

The Valley of the Assassins and the fortress of Alamut was described by Marco Polo, who passed through that area in 1271:

In a beautiful valley, enclosed between two mountains, he had formed a luxurious garden stored with every delicious fruit and every fragrant shrub that could be procured. Palaces of various sizes and forms were erected in different parts of the grounds, ornamented with works of gold, with paintings and with furniture of rich silks. By means of small conduits contained in these buildings, streams of wine, milk, honey and some of pure water were seen to flow in every direction. The inhabitants of these places were elegant and beautiful damsels, accomplished in the arts of singing, playing upon all sorts of musical instruments, dancing, and especially those of dalliance and amorous allurement. Clothed in rich dresses, they were seen continually sporting and amusing themselves in the garden and pavilions, their female guardians being confined within doors and never allowed to appear. The object which the chief had in view in forming a garden of this fascinating kind was this: that Mahomet having promised to those who should obey his will the enjoyments of Paradise, where every species of sensual gratification should be found, in the society of beautiful nymphs, he was desirous of it being understood by his followers that he also was a prophet and a compeer of Mahomet, and had the power of admitting to Paradise such as he should choose to favor. In order that none without his license should find their way into this delicious valley, he caused a strong and inexpungable castle to be erected at the opening to it, through which the entry was by a secret passage.[7]

Hasan-i-Sabah began to attract young men between the ages of twelve and twenty from the surrounding countryside, particularly those whom he marked out as possible material for the production of killers. Says Polo:

[Every] day he held court, a reception at which he spoke of the delights of Paradise, and at certain times he caused draughts of a soporific nature to be administered to ten or a dozen youths, and when half dead with sleep he had them conveyed to the several palaces and apartments of the garden. Upon awakening from this state of lethargy their senses were struck by all the delightful objects, and each perceiving himself surrounded by lovely damsels, singing, playing, and attracting his regards by the most fascinating caresses, serving him also with delicious foods and exquisite wines, until, intoxicated with excess and enjoyment, amidst actual rivers of milk and wine, he believed himself assuredly in Paradise, and felt an unwillingness to relinquish its delights. When four or five days had thus been passed, they were thrown once more into a state of somnolence, and carried out of the garden. Upon being carried to his presence, and questioned by him as to where they had been, their answer was 'in Paradise, through the favor of your highness'; and then, before the whole court who listened to them with eager astonishment and curiosity, they gave a circumstantial account of the scenes to which they had been witnesses. The chief thereupon addressing them said: 'We have the assurance of our Prophet that he who defends his Lord shall inherit Paradise, and if you show yourselves to be devoted to the obedience of my orders, that happy lot awaits you.'[7]

Suicide was reportedly attempted by some in an effort to quickly gain paradise, but the survivors were early told that only death in the obedience of Hasan-i-Sabah's orders could give the key to Paradise. In the eleventh century it was not only credulous Persian peasants who would have believed such things were true. Even among more sophisticated people the reality of the gardens and tours of paradise were completely accepted. True, a good many Sufis preached that the garden was allegorical—but that still left more than a few people who believed that they could trust the evidence of their senses.

The Arab book *Art of Imposture,* by Abdel-Rahman of Damascus, gives away another trick of Hasan-i-Sabah's. He had a deep, narrow pit sunk into the floor of his audience-chamber. One of his disciples stood in this, in such a way that his head and neck alone were visible above the floor. Around the neck was placed a circular dish in two pieces which fitted together, with a hole in the middle. This gave the impression that there was a severed head on a metal plate standing on the floor. In order to make the scene more plausible (if that is the word) Hasan-i-Sabah had some fresh blood poured around the head, on the plate.

Now certain recruits were brought in. "Tell them," commanded the chief, "what thou hast seen." The disciple then described the delights of Paradise. "You have seen the head of a man who died, whom you all knew. I have reanimated him to speak with his own tongue."

After the strange audience with the talking head, when the suitably impressed young Assassins had left, Hasan-i-Sabah had the plate removed from around the head of his accomplice and then with a swift swipe of his sword he beheaded the unwitting accomplice for real, and the severed head would be stuck for some time somewhere that the faithful would see it. This conjuring trick (plus murder) had the desired effect, increasing the enthusiasm for martyrdom to the required degree.

The Assassins fought in the Crusades on whichever side served their purposes. At the same time they continued the struggle against the Persians. The son and successor of Nizam-ul-Mulk was laid low by an Assassin dagger. The new Sultan, who had succeeded his father, Malik Shah, gained power over most of his territories and was marching against the fortress of Alamut. One morning, however, he awoke with an Assassin weapon stuck neatly into the ground near his head. Within it was a note, warning him to call off the proposed siege of Alamut. Out of fear, he came to terms with the Assassins, powerful ruler though he undoubtedly was. They had what amounted to a free hand, in exchange for a pact by which they promised to reduce their military power.

Hasan-i-Sabah lived for thirty-four years after his acquisition of Alamut. On only two occasions was he said to have ever left his room. From his room in Alamut he ruled an invisible empire that stretched across the Middle East. Toward the end, he seemed to realize that death was almost upon him, and calmly began to make plans for the perpetual continuance of the Order of the Assassins.

Hasan-i-Sabah was able to create an independent state based on the Nizari Ismaili theology that was to last for 166 years with Alamut as its capital. Hasan-i-Sabah was succeeded by his most trusted general, Buzurgumid, who reigned as lord of the Nizari Assassins from 1124 to 1138. A few days before his death, Buzurgumid passed the leadership to his son Muhammad, a conservative man who did little to build the Nizari-Assassin empire. He reigned from 1138 to 1162 and then passed the leadership to his son Hasan II.

Hasan II was to reign from 1162 to 1166, and went so far as to declare himself (in the "proclamation of Qiyama") the hidden Ismaili Imam, meaning that he was a form of divinity who communicated directly with God. Hasan II was to become a dangerous heretic to Sunni and Shiite Muslims who rejected him as the Hidden Imam. Hasan II was murdered a year and a half after his proclamation, stabbed to death by his brother-in-law, who vigorously refused to follow the Qiyama teachings and apparently hoped to return the Nizari Ismailis back to their previous theology.

Hasan II was succeeded by his son, Muhammad II who reigned as lord of Alamut from 1166 to 1210. He was succeeded by his son Hasan III, whose mother was a Sunni Muslim. Hasan III caused the Assassins to embrace Sunni orthodoxy. He was succeeded by his nine-year-old son who, as he grew older, quietly returned to the Shiite side of Islam. Muhammad III reigned from 1221 to 1255 and was succeeded by the last of the Nizari-Assassin chiefs, his son Khurshah who reigned from 1255 to the final destruction of Alamut by the invading Mongol hordes. The power of the Nizari Assassins had been broken once and for all.

As James Wasserman says very eloquently in his book, Hasan-i-Sabah has gone down in history as the man who turned assassination into an art form—"maximizing the political benefit of minimum loss of life and offering a more humane method of resolving political differences than the carnage and suffering of the traditional battlefield. Assassination has the unusual effect of entering directly into the halls of power and touching the decision makers themselves rather than the average citizen, the age-old victim of the political adventurism of his leaders."[34]

§§§

The Templars and the Assassins

When the Templars joined the other crusaders in 1129, led by Hugh de Paynes, little did they know that at a future time they would become allied with the Assassins in an attempt to take Damascus from the Seljuk Turks. From this alliance some historians have theorized that the Templars and Assassins were somehow part of the same secret society. This network of mystic secret societies is said to encompass such seemingly diverse groups as Sufis, the Rosicrucians, the Knights Templar and the Masons, Taoists, Tantric Buddhists, Himalayan yogis, Druids and Celtic monks.

Hasan-i-Sabah had sent ambassadors from Alamut in northwestern Persia into Syria and Iraq around the year 1090. They were to form a branch of the Nizari Assassins known today as the Syrian Assassins. Syria at that time was a battleground between two Moslem Empires, the Fatamid Empire based in Cairo and the Seljuk Turk empire based in Constantinople-Istanbul. Meanwhile, a Christian kingdom had been created by the crusaders in Palestine with Jerusalem as its capital.

It was the Syrian Assassins, who still took orders from Alamut, that the Templars and other crusaders were to come into contact with. It was also the chief of the Syrian Assassins who was to become known as the "Old Man of the Mountain," whom we will visit in the next section. Throughout the Mediterranean, troubadours would sing of this man, the Old Man of the Mountain who was feared by kings and governors throughout North Africa and the Middle East.

According to James Wasserman in his carefully-researched book *The Templars and the Assassins*,[34] the Templars' first contact with the Nizari Assassins took place when the crusaders attacked an Assassin-held castle in September 1106. Tancred, a Christian prince of Antioch in northern Syria (now Turkey) attacked the

Assassins at the castle of Apace outside of Aleppo. The Christians defeated the Assassins and leveled a tribute against them. Four years later in 1110, Tancred gained more territory from the Nizari Assassins. Nevertheless, Nizari Assassins helped the Seljuk Turks expel crusader troops from a number of strongholds in northern Syria and Turkey.[34]

However, when the new Seljuk sultan Muhammad Tapar came to power in 1113, he began a campaign against Ismaili Muslims in northern Syria, including the Nizari Assassins. He sanctioned the destruction of the Nizari community in Aleppo and caused several hundred of them to be imprisoned or murdered and their properties seized. All Nizaris and other Ismaili Muslims were expelled from the area by decree in 1124.

The crusaders continued to expand in the area, consolidating power for their Christian kingdom centered on Jerusalem. The Nizari Assassins moved to the Damascus area under their Syrian chief or *dai*, Bahrain. Bahrain's soldiers provided critically needed military support against the Franks and he was given the Syrian frontier fortress of Baniyas by the Seljuk Vizier Tughitigin.

Bahrain began to fortify the castle at Baniyas; he sent out missionaries, and conducted military operations throughout Syria against the Egyptian Fatimids and the crusaders. During one of these operations he was killed; his head and hands were taken to Cairo, where they earned a large reward from the Fatimid caliph.

In 1128, Tughitigin died and an anti-Ismaili wave arose in Damascus reminiscent of the persecutions in Aleppo. Tughitigin's son Buri began the attack by murdering the Vizier al-Mazdaqani and publicly exposing his severed head. This gave the signal for a general massacre of some six thousand Ismailis in Damascus. Rumor spread that the Assassins had made an alliance with the Franks to betray Damascus in return for Tyre. While this was untrue, Bahrain's successor, al-Ajami, had written to Baldwin II, king of Jerusalem, with an offer to surrender the Baniyas fortress in exchange for safe haven from his Sunni persecutors. Al-Ajami died in exile among the Franks in 1130. In 1131, two Nizaris sent from Alamut assassinated Buri. It was during this period (the 1130s) that Buzurgumid succeeded in assassinating the Fatimid caliph al-Amir, virtually ending Mustalian Ismailism in Syria.

After the Nizari Assassins failed to secure a base in the urban centers of Aleppo and Damascus, they purchased the important fortress of Qadamus in the Jabal Bahra mountain range in 1132 under the leadership of Abul Fath. Over the next seven years they successfully acquired eight or ten more castles in the area.

In 1140 they took the important fortress at Masyaf. In 1142, the Hospitallers received the nearby castle of Krak des Chevaliers, becoming hostile neighbors of the Nizari Assassins. In 1149, the Assassins cooperated with Raymond of Antioch in an unsuccessful battle against the Turkish Zangids, during which both Raymond and the Assassin leader Alf ibn Wafa were killed. The alliance with Raymond was motivated by Nizari-Assassin perception of Raymond's strength against the Zangids, who had just taken Aleppo.

The Zangids were enemies of the crusaders, but they were also mortal enemies of the Nizari Assassins. The Zangids championed the Sunni cause to the point of declaring all Shiites, including Ismailis to be heretics who must be eliminated. The Zangids were also frequently allied with the Seljuk Turks.

With shifting alliances, treaties and tributes being paid, the Nizari Assassins ended up paying an annual tribute to the Templars. In 1151, the Assassins battled the crusaders over the town of Maniqa, and in 1152 they assassinated their first Templar victim, Count Raymond II of Tripoli. This assassination of a leading Templar governor led to a Templar attack against the Nizari Assassins at one of the Syrian castles and the imposition of an annual tribute of some two thousand gold pieces payable to the Templars.

One of the most important Christian victims of the Nizari Assassins was Conrad of Montferrat, the Latin king of Jerusalem who was murdered in 1192. He was succeeded by Henry, Count of Champagne, the nephew of King Richard the Lionhearted. Henry, now the Latin king of Jerusalem, was approached by the Nizari Assassins to negotiate a truce.

Arkon Daraul describes Henry's travels to the headquarters of the Syrian-Assassin chief, known as the Old Man of the Mountain, in 1194:

The chief sent some persons to salute him and beg that, on his return he would stop at and partake of the hospitality of the castle. The Count accepted the invitation. As he returned, the Dai-el-Kebir (Great Missionary) advanced to meet him, showed him every mark of honor, and let him view his castle and fortresses. Having passed through several, they came at length to one of the towers which rose to an exceeding height. On each tower stood two sentinels clad in white. 'These,' said the Chief, pointing to them, 'obey me far better than the subjects of your Christians obey their lords'; and at a given signal two of them flung themselves down, and were dashed to pieces. 'If you wish,' said he to the astonished Count, 'all my white ones shall do the same.' The benevolent Count shrank from the proposal, and candidly avowed that no Christian prince could presume to look for such obedience from his subjects. When he was departing, with many valuable presents, the Chief said to him meaningfully, 'By means of these trusty servants I get rid of the enemies of our society.'[7]

§§§

The Old Man of the Mountain—Rashid al-Din Sinan

The most famous of the Syrian Assassin chiefs, the quintessential Old Man of the Mountain, was Sinan ibn Salman ibn Muhammad, also known as Rashid al-Din Sinan (r. 1162-92). Sinan is another legendary Nizari spiritual leader. Like Hasan-i-Sabah, the revolutionary founder, and his spiritual heir, the short-lived Hasan II who burst the chains of Islamic orthodoxy, Sinan was a charismatic and powerful man who changed history.

Sinan was born near Basra in southern Iraq to a well-to-do family and fled to Alamut while a young man after an argument with his brothers. At Alamut he befriended the young Hasan II, his fellow student. Upon Hasan's accession to the position of Imam, he sent Sinan to Syria. Sinan traveled to Kahf, where he worked quietly to develop a following. When the aged chief dai Abu Muhammad died, Sinan succeeded to the leadership of the Syrian Assassins after a brief succession struggle was resolved by decrees from Hasan II.

The Syrian Nizari Assassins were in a precarious position because of the ongoing crusader wars, which pitted the crusaders against the warring Egyptian Fatamid empire and the Seljuk Turks. If the situation could be manipulated correctly, the Nizari Assassins could possibly expand their own territory out of Alamut into northern Syria as far as Damascus. This was largely a Sunni Muslim area, but with the help of the Knights Templar and other crusaders, they might accomplish their goal. Alliances would need to be formed and tribute would have to be paid.

Saladin (1137-1193) became Sultan of Egypt after having been a general of the conquering Syrian army. This ended the Fatimid dynasty and Saladin would now start the Ayyubite dynasty. Originally from Kurdistan, Saladin now controlled and expanded an empire that stretched from Yemen to Tunisia. His main battlefronts were the crusader kingdom in Jerusalem and the Nizari Assassins of Persia and Iraq. In 1171 Saladin declared Sunnism the religion of Egypt. In the throes of Sunni Muslim puritanism against Shiites and Ismaili Muslims, Sunni vigilante groups, called Nubuwwiyya, were organized and roamed the countryside in search of groups they could attack.

The Old Man of the Mountain, Sinan, thus faced a complex series of challenges. On the diplomatic front, he needed to form skillful alliances so that he did not fall victim to a unified Muslim attack. This, at first, inclined him to support the Seljuk Turk leader Nur al-Din as the lesser of two evils, Saladin being recognized as the more implacable foe. He also needed to cultivate a successful relationship with the crusaders, with whom he, as a Muslim, was technically at war. The Assassins were already paying a tribute to the Knights Templar by the time Sinan took charge. He must have considered this money a reasonable investment to prevent open warfare. Finally, he needed to establish and fortify more defensive strongholds to maintain himself against attack from any of several different potential sources of aggression, including hostile neighbors.

In 1173, Sinan sent an ambassador to King Amalric I of Jerusalem proposing an alliance. He asked for a relaxation of the tribute he was paying to the Templars as his only condition. It is commonly believed that Amalric agreed to Sinan's terms, and that the Assassin ambassador was murdered by the Templars during his return journey in order to prevent their losing the tribute. It is also generally accepted that the murder prevented further progress regarding an alliance, despite Amalric's apology to Sinan and his imprisonment of the responsible knight. The potential for additional discussion was obviated by Amalric's death soon after in 1174.

Archbishop William of Tyre, a contemporary historian, wrote that Sinan had expressed, through his ambassador, the willingness of the Assassins to convert to Christianity in furtherance of such an alliance. Farhad Daftary suggests this was a gross misinterpretation of Sinan's offer. The sophistication of Sinan's theological interests and his ecumenical viewpoint would naturally incline him to learn more of the religious doctrines of his potential allies, but not to convert. The Qiyama doctrine, proclaimed Sinan's friend Hasan II, was misinterpreted by more than one medieval historian as a distorted acceptance of Christianity and a rejection of Islam, which raises the important point that certain elements of the Qiyama doctrine are not incompatible with aspects of Christianity. The revolutionary stance taken by Jesus in rejecting much of the outward observance of Jewish law certainly finds an echo in the Qiyama teachings.

It is believed that at the time of his death in 1174, Nur al-Din was planning an expedition against the Ismailis in retaliation for their suspected arson of a mosque in Aleppo. Upon learning of Nur al-Din's passing, Saladin proclaimed his independence from the Zangid dynasty and established himself as the first ruler of the Ayyubid dynasty. Saladin was a threat to Nur al-Din's young son and heir, al-Malik al-Salih. The regent who ruled in the boy's name sought the help of Sinan in stopping Saladin by assassination. One reason for Sinan's cooperation was the hatred the Assassins had developed for Saladin. In 1174-75, Nubuwwiyya vigilantes had raided two Ismaili centers, killing some thirteen thousand people. Saladin was passing by at the time. He learned of the massacres and took advantage of the situation to attack other Nizari strongholds before moving on.

Sinan sent some fidayi against the Ayyubid sultan in 1175. A local ruler with whom Saladin was visiting recognized them and thwarted the attempt. A second effort was made in 1176. Saladin was slightly wounded, but he was saved by his own quick reaction and the chain mail armor he wore at all times.

Then, a curious thing happened in August 1176. Saladin attacked the castle of Masyaf and laid siege to this central Nizari-Assassin fortress. Without warning, he ended the siege and departed. Various explanations have been given, including the following reliable account from Sinan's biographer. One day a messenger from Sinan approached Saladin. He stated that the message was personal and must be delivered only in complete privacy. Saladin progressively emptied his court until only two Mameluke attendants were left.

Sinan's messenger asked Saladin why he would not order the Mamelukes to depart so he could deliver his message in private. Saladin replied, "I regard these two as my own sons. They and I are one."

The messenger then turned to the Mameluke guards and said, "If I ordered you in the name of my Master to kill this Sultan, would you do so?" They drew their swords together and replied, "Command us as you wish."

Saladin, realizing the danger he was continually in at the hands of the Nizari Assassins, apparently decided to call a truce with the group. This incident seems to have henceforth allied the two leaders and there are no more records of conflict between Saladin and Sinan, the Old Man of the Mountain.[34]

One important consequence of their alliance may have been the assassination of the Frankish king of Jerusalem, Conrad of Montferrat, in 1192, shortly before Sinan's own death. Two fidayi disguised as Christian monks were responsible for Conrad's murder.

This assassination was a major blow to the crusaders, and a number of accusations have been made about it, including the suggestion that it was done at the behest of King Richard I, "the Lionhearted." Sinan would have had plenty of motives himself, including possible instructions from Saladin. Conrad had recently seized a Nizari ship and its cargo, and his murder of the crew was probably sufficient motivation for Sinan to order the assassination.

In any case, the death of Conrad aided Saladin in his efforts to conclude a truce with Richard. The Nizaris were included in the treaty at Saladin's request. Sinan and Saladin died within months of each other, while Richard left the Holy Land for Europe.

Sinan was a nearly mythological religious figure among the Syrian Nizaris. He had no bodyguards to protect him, ruling through the sheer force of his personality. He traveled from fortress to fortress, with no permanent base or established bureaucracy. Because of his continual movement, the network of Ismaili fortresses was both close-knit and ever alert. William of Tyre, writing during Sinan's reign, estimates that there were some sixty thousand Syrian followers of the Assassin chief. Sinan was described by another historian as a tender and gentle ruler.

The Nizari Assassins were a mystical sect, with strong ties to the ancient traditions of the Sufis, Zoroastrians, Hindus and Buddhists. Afghanistan was a Buddhist kingdom during the time of Alexander the Great and remained one until the advent of Islam.

Sinan, whose education was more along the lines of the Sufis and the Hindu mystics of the Indian subcontinent, was reputed to be an advanced astrologer and alchemist, as well as adept in the arts of magic, telepathy, and clairvoyance. He was never seen to eat or drink and was credited with healing powers. He was twice said to have prevented great rocks from crushing people by the use of psychokinetic powers. He was reputed to have psychically repelled an attack by Saladin's soldiers by holding them immobile and beyond striking distance of the Nizari-Assassin troops on the battlefield.

Sinan's courtesy was as much a part of his legend as his clairvoyance. One anecdote tells of his visit to a village in which he was honored by the local leader, who brought him a covered plate of food specially prepared by his wife. Sinan gently had the plate moved to one side. The official was disappointed at this apparent rejection of his hospitality and asked Sinan the reason. Sinan quietly took him aside and explained that, in her excitement, the village chief's wife had forgotten to properly clean the insides of the chickens. If the plate were uncovered, the chief would have been embarrassed before his people. On checking beneath the cover, he discovered that Sinan was correct.

Sinan subscribed to Hasan II's Qiyama proclamation and was charged with promulgating the doctrine in Syria. He held the Qiyama feast of Ramadan soon after his arrival. He seems to have accepted Hasan II as a legitimate spiritual Master; however, he was unwilling or unable to transfer his allegiance to Hasan II's son and heir, Muhammad II. Thus the Syrian Assassins became independent of Alamut from the death of Hasan II until the death of Sinan. This was unacceptable to Muhammad II, who sent several different groups of fidayi against Sinan. But Sinan is said to have either killed or won over each of his would-be Persian assailants.

Many Syrians believed Sinan was the divinely ordained Imam or at least his *hujja.* Some even regarded him as an incarnate deity. One story has him discovered by his disciples in a nocturnal conversation with a green bird glowing with light. Sinan explained that this was the soul of Hasan II come to request the help of the Syrian Imam. Sinan was honored beyond precedent by a Syrian shrine built in his honor.[34]

§§§

The Secret Alliance Between the Templars and the Assassins

It has been thought by some historians that the Templars and the Nizari Assassins had a number of agreements that encompassed tribute paid to the Templars and assassinations made on their behalf.

Richard the Lionheart was at one time accused of having asked the 'Lord of the Mountain' to have Conrad of Montferrat killed; a plot which was carried out thus: "Two assassins allowed themselves to be baptized and placing themselves beside him, seemed intent only on praying. But the favorable opportunity presented itself; they stabbed him and one took refuge in the church. But hearing that the prince had been carried off still alive, he again forced himself into Montferrat's presence, and stabbed him a second time; and then expired, without a complaint, amidst refined tortures."[34]

The famous question of the three thousand gold pieces paid

by the Syrian branch of the Assassins to the Templars is another matter which has never been settled. One opinion holds that this money was given as a tribute to the Christians; the other, that it was a secret allowance from the larger to the smaller organization. Those who think that the Assassins were fanatical Moslems, and therefore would not form any alliance with those who to them were infidels, should be reminded that to the followers of the Old Man of the Mountain only he was right, and the Saracens who were fighting the Holy War for Allah against the crusaders were as bad as anyone else who did not accept the Assassin doctrine.

Grave charges against the Templars during the Crusades included the allegation that they were fighting for themselves alone. More than one historical incident bears this out. The Christians had besieged the town of Ascalon in 1153, and were engaged upon burning down the walls with large piles of inflammable materials. Part of the wall fell after a whole night of this burning. The Christian army was about to enter, when the Master of the Temple (Bernard de Tremelai) claimed the right to take the town himself. This was because the first contingent into a conquered town had the whole spoils. As it happened, the garrison rallied and killed the Templars, closing the breach.

Says Arkon Daraul in his *A History of Secret Socities,* "There seem good grounds for believing that the power which they had gained caused the Templars to devote their efforts as much to their own Order's welfare as to the cause of the Cross, in spite of their tremendous sacrifices for that cause. Having no loyalty to any territorial chief, they obeyed their Master alone, and hence no softening political pressure could be put upon them. This might well have led to an idea that they were an invisible super-state; and this does show some similarity with the invisible empire of the Assassins. If none can deny their bravery, their high-handedness and exclusivity in less than a hundred and fifty years after their founding gave them the reputation of considering themselves almost a law unto themselves."[7]

§§§

The End of the Templars and the Assassins in the Holy Land

In 1184 an incident occurred which inspired a great deal of distrust of the Templar Order. The English knight Robert of St. Albans left the Templars, became a Moslem and led an army for Saladin against Jerusalem, then in the hands of the crusaders. The charge was then brought against the Templars that they were secret Mussulmans or allies of the Saracens.

Two years before, Saladin, had made a pact with the Assassins that they would give him a free hand to continue his Holy War against the crusaders. On July 1st, 1187, he captured Tiberias. He attacked nearby Hittin on Friday, July 3rd. Thirty thousand crusaders were captured in the decisive battle of Hittin, including the King of Jerusalem. This effectively ended the power of the crusaders.

No Templar is mentioned in the detailed Arab account as asking for mercy on religious or other grounds, although all knew that Saladin had issued a war-cry: "Come to death, Templars!" The Grand Master, Gerard of Ridefort, and several other knights were among those taken. Saladin offered them their lives if they would see the light of the True Faith. None accepted, and all these knights were beheaded except the Templar Grand Master. A non-Templar, Reginald of Chatillon, tried to invoke the sacred code of Arab hospitality, and other crusaders claimed that they were Moslems, and were spared— none of them Templars.

Reginald and the Templars collectively were sentenced to death for breaking the truce and the "war crime" of killing unarmed pilgrims to Mecca. However, Templar Grand Master Gerard of Ridefort was held in prison by Saladin to be ransomed. Ridefort was ransomed four years later, in July of 1188, a year after Saladin had captured Jerusalem. Ironically, he was killed a year later as the Templars sieged the port of Acre. Though they won that battle, their power in Holy Land would never be the same.

The Battle of Hittin was the turning point for the Crusaders and spelt the end of real Western power in Palestine for over seven hundred years, although it did stimulate the unsuccessful Third Crusade. Although the Templars—and some other crusaders— were still in the Holy Land, they had lost almost all of their possessions there. But in the West lay the real seat of their power. At this time their European possessions numbered over seven thousand estates and foundations. Although principally concentrated in France and England, they had extensive properties in Portugal, Castile, Leon, Scotland, Ireland, Germany, Italy and Sicily.

When Jerusalem was lost, their headquarters were transferred to Paris. This building, like all their branch churches, was known as the Temple. It was here that the French King Philip the Fair took refuge in 1306, to escape a civil commotion. It is said that this visit gave him his first insight into the real wealth of the order: the fabulous treasures which his hosts showed him giving the bankrupt monarch the idea to plunder the knights on the pretext that they were dominated by heresies.

The Assassins' power was finally checked by the Mongol invaders, who made a special point of destroying them in Persia, including burning their villages. The Mongols had arrived in the vicinity of Persia in 1219 and it is thought that the Hasan III of Alamut was the first Islamic ruler to send ambassadors to the Mongol Khan, arranging some sort of truce.

Hasan III died in 1221 and was succeeded by Muhammad III, who ruled Alamut under an uneasy truce with the Mongols until his death in 1255. By 1240 the Mongol assault had reached as far as western Persia. In 1248 an Ismaili truce delegation which included Nizari Assassins was turned away by the Mongol leaders.

In 1254 the Mongol chief Mangu recieved William of Rudrick, a Franciscan friar and ambassador for Louis IX. The French monarch had wished to enlist the Mongols to support the Christian armies in the Seventh Crusade.

During this meeting William learned that Mangu was in fear for his life because he had heard that as many as 40 Nizari Assassins in various disguises had been sent after him to retaliate for his campaign against them and western Persia which had started a few years before.

Mongol armies began arriving in the Alamut area in the same year, and attacks on the Nizari Assassins and their stronghold of Alamut commenced. In 1255 Muhammed III was murdered by his homosexual lover, and his son, Khurshah succeeded him. In 1256 the Mongols under the command of Huelgu Khan began their final siege of Alamut. Huelgu sent a messenger to Khurshah that if he would surrender and destroy his fortress, he would be allowed to live.

Khurshah asked for one year before presenting himself to the Khan. The Mongol army left but returned in November of 1256, giving Khurshah five days to surrender, though the agreed upon year was not complete. On November 19 Khurshah entered the Mongol camp with his family, entourage and treasure. The remaining Nizari Assassins in Alamut refused to surrender and were eventually killed by the Mongols.

The Mongols were amazed by the fortress of Alamut. Water ran through rock-cut channels and enormous water tanks were carved out of solid rock. There were great food stores and the castle had been designed to withstand a long siege. With great difficulty, the Mongols destroyed much of the fortress.

A large library was kept at Alamut and the Nizari Assassins were said to be great devotees of knowledge and books. The Persian historian Juvayni accompanied Huelgu's army and personally oversaw the burning of the entire library.

The Mongols moved on to other Assassin strongholds and destroyed them. Khurshah wrote to the Syrian Assassins and commanded that they surrender to the Mongols. Khurshah also asked for an audience with the great Mangu Khan himself and embarked in 1257 on the long journey to Karakoram in central Mongolia, accompanied by Mongol troops. When he arrived, the Khan refused to see him because the Syrian Assassins were still holding out against the Mongols. On the return journey to Persia Khurshah was beaten and then stabbed to death by his Mongol guards.

The Mongols began to exterminate all relatives of Khurshah, and all Ismailis in general. Khurshah's son Shams al-Din Muhammad was whisked away to safety, and the current Aga Khan of the Ismailis is descended from this surviving son.

The Syrian branch of the Assassins was wiped out a few years later at the hands of the Baybars, the army under control of the Mameluke Sultan of Egypt who had decisively defeated the Mongols in Syria in 1261 and exterminated the last of the Assassin strongholds.

And so we have met the group that brought the words "assassin" and "assassination" to European languages. The concept of "decapitation" strikes on the enemy entered into the mind of the Templars and other crusaders. Large armies could still meet on the battlefield while the kings and governors watched the slaughter from hilltops, but another way to wage war was to take the death and destruction directly to the leaders themselves. While the masses of humanity are manipulated and physically forced to march into battle as cannon fodder, the ruling elite have been content to watch the violent consequences of their decisions and declarations, but careful to keep themselves and their family out of the fighting.

The Nizari Assassins changed all that and brought the wars and power struggles directly to the leaders. The Templars—soon to be betrayed and hunted to the death much like the Nizari Assassins—would learn a few things from Hasan-i-Sabah. They would strike directly at the leaders of armies who fought against them, and take their battle against the Pope and the Vatican to the high seas. But if the Templars were to ultimately control the sea, they would need naval bases and seacharts. And because of the access to the libraries of the Middle East, the Templars were able to acquire copies of the maps of the ancient sea kings.

The crusaders fought Egyptian and Turkish armies.

The Assassins in this Victorian portrait.

The crusaders the Saracens in this early painting.

The Knights Hospitallers at Rhodes, shortly before moving
to Malta and changing their name to the Knights of Malta.

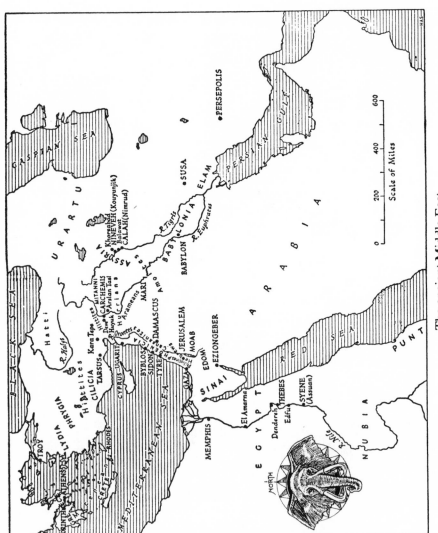

The ancient Middle East.

Seljuks in flight from Crusaders, from the Abbey of St Denys

An Eastern Knight, from the Apocalypse of St Severo

Siege towers were constructed some distance from the city walls

Saracens parley with Crusaders in their camp, from *Chroniques de St Denis*

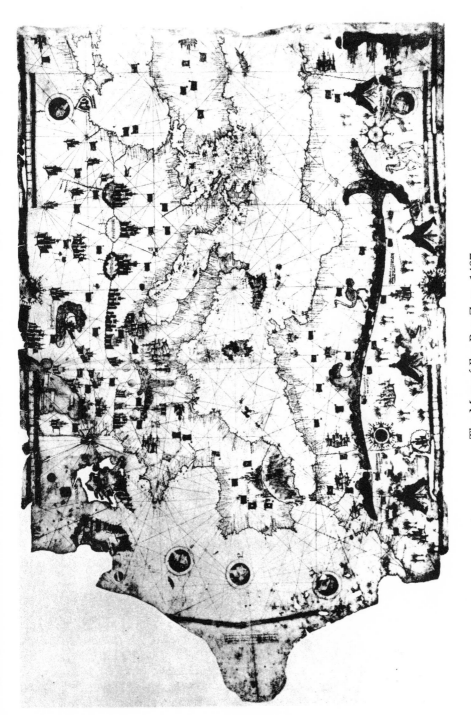

The Map of Ibn Ben Zara, 1487.

Mercator's World Map of 1538.

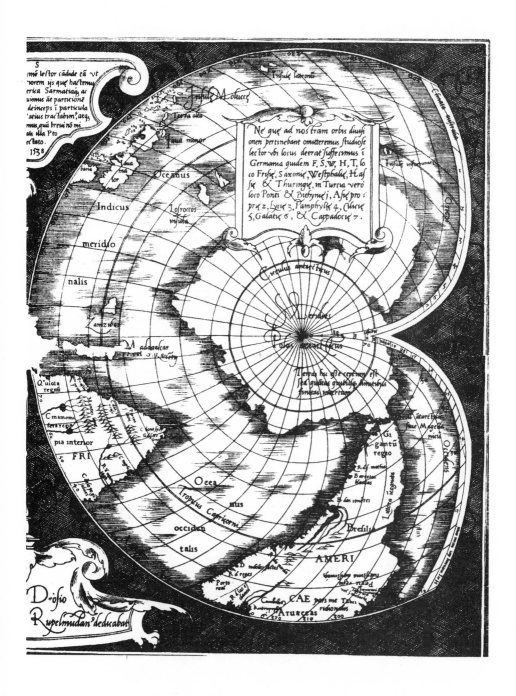

The World Map of Idrisi.

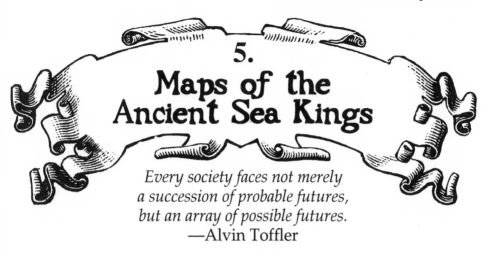

5.
Maps of the Ancient Sea Kings

*Every society faces not merely
a succession of probable futures,
but an array of possible futures.*
—Alvin Toffler

*To succeed it is necessary to accept
the world as it is and rise above it.*
—Michael Korda

It has been suggested by some historians that the kings of ancient seafaring nations were the navigators, the highly educated men who could read the stars and navigate a vessel across the wine dark sea. As we discussed in Chapter One, seafaring, navigation and the adventurous sailing voyages of exploration have been going on for thousands of years.

While modern isolationist historians maintain that oceans are barriers to world travel, the modern diffusionist maintains that oceans are highways, not barriers. It is much safer in the long run for a ship to carry passengers and cargo for thousands of miles along a coastline than to walk that same coastline, paying taxes and battling hostile tribes or bands of robbers along the way. While some ships may be lost in storms or run aground, and others may be attacked by pirates, most would make it across the waves.

For the Templars, the twin defeat of having lost Palestine and being suppressed by the Vatican and the French king, gave them some heavy food for thought. They had lost their status with the Holy Roman Empire, and their very lands and other possessions were taken away from them. Many of their leaders, including the Grand Master Jacques de Molay, were captured, tortured and executed.

The Templars were in a reduced state, indeed, but they apparently still had the cache they guarded, and a magnificent fleet of ships that once flew a white flag with a red cross on it (this flag is still in use today as the Cross of St. George and flies today as the flag of England, which is not the commonly known Union Jack). This fleet could no longer make port in France, Spain, Italy, Ireland, England and many other countries. There were few places it could go: Sicily, Portugal, Scotland and Norway, among a few others.

The Templars needed a refuge, and they were essentially at war with the Vatican and the King of France (as well as other Holy Roman Kings). As stated in the last chapter, the Templar Fleet was to split into three main sections: the southern portion of the fleet was to stay in the Mediterranean and become pirates, flying the Jolly Roger of the Norman kings of Sicily. A second contingent of the fleet was based in Portugal to become the navy of King Henry the Navigator. The northern portion of the fleet made it to Scotland where many of the Templars actually built homes and married.

The Portuguese and Scottish contingents were apparently in possession of a number of ancient maps that were to help them in explorations into the Atlantic— explorations that may have been to find new lands to which the Templars and the Holy Blood-Holy Grail could be safely moved, away from their powerful persecutors. These maps showed is-

lands and continents beyond the known world of Europe of the time. They showed islands in the Atlantic, and even parts of North and South America. There were stories from the Norsemen of habitable lands west of Ireland and Scotland. Iceland was first settled by the Norse around the year 850 AD, and Irish monks had arrived there even before that. Perhaps the Templars would also find a new home.

§§§

The Sudden Appearance of Highly Accurate Ancient Maps

Canadian author and historian Michael Bradley (*Holy Grail Across the Atlantic, The Columbus Conspiracy*) says that the Templars had knowledge of land across the Atlantic. As we have noted, according to Bradley, there were three major repercussions of the Templar dispersion: an upsurge in piracy; the beginning of transatlantic probes by ships from England and Portugal; and the flooding of European ports with sea-charts of inexplicable accuracy.

Bradley says that this last item is the most important of the repercussions of the Templar dispersion. Not only did Templars apparently instigate transatlantic probes just as soon as they arrived in Portugal and Scotland, but all of Europe suddenly seemed to be in possession of sea-charts of previously unequaled quality. These charts are really the most important result of the Templar dispersal because they accomplished two things. First, they greatly facilitated inter-European trade and commerce by sea, and thus contributed to the further decline of feudalism. Second, the charts really made the European "Age of Discovery" possible because they showed the whole world, and also showed new land across the Atlantic.

Within a generation after Philip IV of France pressured Pope Clement V into disbanding the Templars, maps called "portolans" began to be distributed throughout Europe. One of the earliest portolans is called the Dulcert Portolan of 1339, which appeared just 27 years after the Knights Templar were disbanded. Scholars of navigation have consistently tried to ignore the portolans because of the problems they present. It is accepted that they did exist, but the further implications of these charts have been swept under the carpet. Only two experts have truly grappled with the mystery of the portolans, but academia as a whole has turned a blind eye toward their conclusions.

Stated simply, the mystery of the portolans is not so much that they appeared so suddenly in 14th-century Europe but that they are incredibly precise. It will be recalled that the barrier to medieval navigation was that longitude, the position east or west of any given point, could not be determined with any accuracy. The key to finding longitude by celestial observation required the measuring of time with extreme accuracy. Clocks of such sophistication simply did not exist in the medieval world, and navigators had to wait until 18th-century technology supplied them.

The mystery of the portolans is that they are accurate in terms of longitude and, although they appeared suddenly in the medieval world, no one person, culture or civilization of the 14th century could have produced them.

The Scandinavian scholar, A. E. Nordenskøld, studied all the portolans he could find in the 1890s. After analyzing hundreds of these charts from scores of European museums, Nordenskøld concluded that all of the portolans had been copied from just one original chart. This original chart was highly accurate in terms of both latitude and longitude. In fact, it was more accurate than some maps made in Nordenskøld's own day. Other portolans replicated this accuracy to a greater or lesser extent depending on the care taken in the copying—but one and all were much more accurate than maps made by Ptolemy, and far more accurate than products known to have derived from the medieval period.

§§§

The Work of Charles Hapgood

A modern expert, Prof. Charles Hapgood of Keene State Teacher's College in New Hampshire, took up, in the 1950s and 1960s, where Nordenskøld left off in the 1890s. Hapgood arranged to have his analyses checked by the Cartographic Section of the U.S. Air Force's Strategic Air Command (8th Reconnaissance Technical Squadron). Like Nordenskøld before him, Hapgood could only conclude that all the portolans seem to have been copied from one original because they all displayed the same peculiarities. Hapgood also concluded that a portolan known as the Ibn Ben Zara chart, dated to 1487, was the best one and had been most carefully copied from the original.

It was probably not the original, but it best replicated the original. Hapgood says of it: "I had been attracted to a study of this portolan because it seemed definitely superior to all other portolan charts I had seen in the fineness of the delineation of the details of the coasts. As I examined the details in comparison with modern maps, I was amazed to see that no islet, however small, seemed too small not to be noted... The grid worked out for the map revealed, indeed, a most amazing accuracy so far as the relative latitudes and longitudes were concerned. Total longitude between the Sea of Azov and Gibraltar was accurate to half a degree."[36]

This is an error of only 30 miles over an east-west distance of over 3000 miles, and is better accuracy than can be claimed by road maps of the 1950s and 1960s.

All of the surviving portolans are centered on the Mediterranean world. They show only Europe from the Atlantic Coast, through the entire Mediterranean, and usually also include all of the Black Sea. Occasionally, as with the Ibn Ben Zara chart, they extend further north to the Sea of Azov. One or two of the portolans even extend eastward to include the Caspian Sea. Always, the longitude is minutely accurate.

A crucial question begs to be asked. Although the surviving portolans depict only Europe, did the source map from which they all derive depict a much larger area? In fact, could the portolans have been copied from just the European part of a larger map that showed the entire world?

This question is crucial indeed. If the portolans represent just the small European portion of some larger world map, or "mappamundi" as academics would say, then the corollary must be that the rest of the world would be depicted as accurately as the small European part. In short, if there was a mappamundi source it would have accurately shown Europe, Asia, Africa and the Americas because there's no good reason to assume that the remainder of the world would have been drawn any more or less accurately than the European portions which survive. There's a subtlety to this situation which should be emphasized. If a source mappamundi existed that showed Europe, Asia, Africa and the Americas then the owners of this priceless map-of-the-world could see clearly that there was land across the Atlantic, but that it was not Asia.

Was there ever such a world map? Did Templars get possession of it? Did they bring copies to Portugal and to Scotland which served as guides for Atlantic voyages in search of a religious haven in a New World, and not in Asia?

§§§

The Mapapamundi or Ancient Maps of the World

Michael Bradley says, "I believe it is justified to assume that it is probable, with this probability verging on certainty, that Templars did possess such a map of the world. The reason for stating this so strongly is simply that modern scholars have found precisely this kind of mappamundi in Middle Eastern archives. Specifically, two very intriguing maps of the world have been found: the Hadji Ahmed Map was discovered in 1860 in what is now Lebanon; then, in 1929 the Piri Re'is Map was discovered in the old Imperial Palace in Constantinople. Before looking at these maps and their unnerving implications, something should be said about the methodology of medieval and early Renaissance cartographers."

Supposing that a mapmaker of the 15th or 16th century possessed a very accurate map that he could copy, he would still be inclined to "improve" on it based on the best contemporary knowledge available to him. Look at the situation from the cartographer's point of view: he may have a map showing the whole world and places that had not yet been discovered, but he had no idea how accurate this map might be. He knew well enough how accurate his own charts were. Not very accurate at all. Why should he assume that his priceless mappamundi was any more reliable than the best maps of his own age? He had no good reason to make any such assumption, and therefore he had every right, and even an obligation, to correct matters to the best of his ability.

For places that were unknown by reason of inaccessibility, such as northern Greenland, for example, or for places depicted on the mappamundi that had not yet been discovered, the mapmaker had no choice but to rely on his source map. But when it came to places that the cartographer thought he knew, like the Atlantic Coast of Europe and the Mediterranean area, he felt an obligation to improve things from the best knowledge available to him. Unfortunately for the medieval or early Renaissance cartographer, and unfortunately for modern scholars, 15th and 16th century knowledge and mapmaking techniques were no match for the accuracy of certain maps that were in the hands of some mapmakers of the 1400s and 1500s.

The mysterious source maps were always accurate, while the mapmaker's own attempts to improve on them with current knowledge always resulted in distortions that stick out like the sore thumb of ignorance. Maps like the Hadji Ahmed Map and the Piri Re'is Map were far more accurate than any known current maps—and showed areas that were supposedly unexplored!

§§§

The Hadji Ahmed Map

First, the Hadji Ahmed Map. This was drawn by an Arab geographer, only obscurely known to history, who operated out of Damascus. It is dated 1559 and it shows the entire world in a somewhat fanciful type of projection that is more art than science and which was typical of Arab chartwork of the mid-16th century. A careful look at this map will show that Hadji Ahmed "improved" the Mediterranean according to Ptolemy, and thus distorted it, and also drew Africa according to the best Portuguese information that he could get, and distorted Africa too in a manner completely typical of the mid-16th century.

But when we look at North America and South America they have an almost modern shape, and could compare well with Mercator's Map of South America drawn 10 years later from contemporary explorers' information. Thankfully, Hadji Ahmed apparently had no access to contemporary maps and charts of the Americas and so he was stuck with simply copying some mysterious mappamundi in his possession.

This unknown source map of Hadji Ahmed was more accurate than the best information available in 1559 and so Hadji Ahmed's Map looks very modern. It shows Baja California, which had not been mapped then. It shows the Northwest Coast of North America, including Alaska, which had not been discovered then. It shows the Hawaiian Islands in the Pacific, which were not discovered until two hundred years later. It shows a sprinkling of islands in the Pacific, a sort of vague and suggestive rendition of the Polynesian Islands, but they had not been discovered yet. It shows Antarctica clearly, and even a suggestion of the Palmer Peninsula, and that had not been discovered then either.

The Far East, insofar as it can be made out in the curious "splitapple" projection used in the map, is distorted but reasonably accurate. But the strange and unnerving thing about this map is the region of Alaska and Asia. The curve of the Aleutian Islands is depicted accurately, but there is no Bering Strait and the whole area is land. This part of the map depicts how the world of that region actually was—but 10,000 years ago! This map shows the "Bering Land Bridge" between Asia and North America, and shows it correctly. Up until the 1958 Geophysical Year, scientists always thought that the Bering Land Bridge had been exactly that, a "bridge," which is to say a rather narrow connecting link between Asia and Alaska. Soundings taken in 1958 proved conclusively that this land connection had not been a narrow bridge at all, but an expanse of land of subcontinental proportions that included all the area north of the curving Aleutian chain and the Alaska panhandle. In short, it had been exactly as shown on Hadji Ahmed's Map.

This fact almost defies speculation. Or, is it just a coincidence? Perhaps a mediocre mapmaker, not knowing how Asia and North America actually terminated, decided to make things easy and simply join them. Hapgood and Bradley both believed that the

Ibn Ben Zara Chart, like all the portolans, and particularly like the best of them, shares a peculiarity that characterizes all portolans: the general accuracy is there but the sea level seems too low.

On the Ibn Ben Zara Chart, most of the islands of the Aegean which currently exist are all shown a bit larger than they are today, while there are some "extra" islands that do not presently exist, but which would exist if the sea level dropped by about 200-300 feet. These islands did exist ten thousand years ago, near the end of the Ice Age when the sea level was exactly 200-300 feet lower than today! Also, on the Ibn Ben Zara Chart, we see that some river deltas, like those of the Nile and the Rhone, are shown significantly smaller than they are today as if the rivers were younger and had just commenced flowing after the ice retreated.

On the Hadji Ahmed Map, we see the same phenomenon replicated in other places besides the Aleutian area. There's a large bump on southern California extending out into the Pacific, and this is the actual location of a bit of continental shelf that was above the sea ten thousand years ago. Looking at northeast North America, we can see an inlet that represents either the Bay of Fundy or the St. Lawrence estuary, but Nova Scotia and Newfoundland are joined together. This would have been the case ten thousand years ago because the Grand Banks of Newfoundland and the George's Bank off Nova Scotia would have been above sea level. Professor Steve Davis, of St. Mary's University in Halifax, Nova Scotia, has just caused a stir among Canadian archeologists with his announcement that human artifacts have been dredged up from the George's Bank by a scallop boat. Davis dates the find to about ten thousand years ago. People inhabited this now-drowned land. Although the Canadian newspapers naturally headlined the discovery with sensational allusions to Atlantis, the artifacts discovered are primitive fish processing utensils very similar to those used by today's Eskimo people and by the vanished Beothuk Indians of Newfoundland.

Since these sea-level problems are common to all the portolans and to the existing mappamundi from which the portolans seem to have been excerpted, are we to believe that the earth was accurately mapped ten thousand years ago, and that a few copies survived into the medieval period?

§§§

The Piri Re'is Map

The Piri Re'is Map found in 1929 at the Topkapi Museum in Istanbul presents an even greater puzzle. It was drawn in 1519, the year that Magellan's expedition set out to circumnavigate the world. But this expedition did not return to Europe until 1521 and so the Piri Re'is Map could not have relied on information derived from this voyage. According to marginal notes presumably made by Piri Re'is himself, his map was based on "the map of Columbus" and on other maps "dating from the time of Alexander the Great." Note that Piri Re'is does not say a map by Columbus, but a map of Columbus'. Piri Re'is was first an Islamic pirate and then a Turkish admiral (he was actually Jewish, as were many in the Moorish and Turkish courts), and he might have been in a position to know, or guess, what sort of map Columbus had and from which mappamundi it had been copied.

In any event, this map caused a stir in both diplomatic and geographic circles because it showed the American continents with incredible accuracy. The problem with this was that the American continents had not yet been explored, or even coasted to any great extent, in 1519. Europeans were just then feeling their way out of the Caribbean. Cortez landed in Mexico the same year this map was drawn, 1519. Pizarro had not yet met the Incas of South America. What, then, could be the source of this map? The American Secretary of State of the time, Henry Stimson, began a flurry of correspondence with Turkish authorities that lasted through much of the 1930s. Stimson urgently requested the Turks to conduct a careful search of all their old archives to see whether any similar maps might come to light. The Turks complied, or said they did, but nothing else like the maps of Hadji Ahmed or Piri Re'is turned up.

The Piri Re'is Map does show the New World coastline with incredible accuracy, but it may not seem that way to the average reader. It was self-evident to cartographic experts of the 1930s, however, because they were able to recognize immediately that the Piri Re'is Map had been drawn according to a very special sort of projection: azimuthal equidistant projection. I have tried to explain this, and to demonstrate it visually, in the accompanying illustrations. A careful look at these illustrations, and a serious reading of the descriptions, will clearly convey the staggering implications of this map. Cartographic experts were, and are, both amazed and puzzled by it.

Professor Charles Hapgood spent a great deal of time analyzing the Piri Re'is Map, and the Strategic Air Command endorsed both his methods and his startling conclusions: the Piri Re'is Map could only have been drawn using data from aerial photography! See Hapgood's *Maps of the Ancient Sea-Kings*[36] for a highly-illustrated and detailed discussion of this map and many others. Analysts of world history owe a great debt to Hapgood for compiling these maps and showing us their amazing contents.

§§§

The Zeno Map of the North

Henry Sinclair, Grand Master of the Templars after they fled France to Scotland, apparently had a remarkable map when he set sail for Nova Scotia in 1398 (more on this in the next chapter). This map comes down to us as "The Zeno Map of the North," and was drawn by a Venetian navigator in Sinclair's service, Antonio Zeno, sometime in the late 1300s.

This map was supposedly the result of a voyage made by Antonio Zeno and his brother Nicolo from Venice in the year 1380. Their explorations supposedly took them to Iceland and Greenland, and perhaps as far as Nova Scotia. They drew a map of the North Atlantic which was subsequently lost for two centuries before it was rediscovered by a descendant in the 1550s.

But it is obvious from a detailed study of this map, which Hapgood conducted and recounts in detail in his book, that Antonio Zeno's map was actually copied from some other highly accurate chart that was drawn on a conic projection. Antonio was not familiar with this projection, which is understandable since it was not "invented" until three centuries after his death, and he also "improved" things from his own knowledge when he could.

Rene Noorbergen agrees that the Zeno brothers could not have been the original mapmakers. The brothers supposedly touched land in Iceland and Greenland, yet their chart very accurately shows longitude and latitude not only for these locations, but also for Norway, Sweden, Denmark, the German Baltic coast, Scotland, and even such little-known landfalls as the Shetland and Faroe islands. The original mapmakers likewise knew the correct lengths of degrees of longitude for the entire North Atlantic; thus, it is very possible that the map, instead of being a product after the fact, was drawn up by the Zeno brothers before their voyage and was used to guide them in their exploration of the northern lands.[30]

Just how ancient the original source maps may have been is indicated by the fact that the Zeno map shows Greenland completely free of ice. Mountains in the interior are depicted, and rivers are drawn flowing to the sea, where in many cases glaciers are found today. Captain Mallery, whose initial work on the Piri Re'is map led him to study other Renaissance charts such as the Zeno brothers', took special note of the flat plain shown stretching the length of the Greenland interior on this map, intersected midway by mountains. These features are not discernible today due to the ice coverage, but their existence was confirmed when the Paul-Emile Victor French Polar Expedition of 1947-49 found precisely such topography from seismic profiles.[30]

As with the revelation that Antarctica at one time was free of ice and perhaps inhabited, we find similar legends of a civilized people who once lived in northern lands which are now buried under thousands of feet of ice: the legends of Thule, Numinor and the Hyperboreans. Egerton Sykes, in his *Dictionary of Non-classical Mythology*, page 20, states his belief that the Norse legend of Fimbelvetr, the "terrible winter" that launched the epic disasters of Ragnarok and destruction of the gods of Valhalla, may reflect a historical fact: the obliteration of a prehistoric civilization in the boreal regions by the Ice Age catastrophe.[30]

It is interesting to speculate that the Zeno brothers map shows what may have been the lost land of Thule, a legendary northern land mentioned by such Greek and Roman historians as Diodorus Siculus (*The Library of History*, 1st century BC), Strabo (*Geography*, 1st century BC) and Procopius (*The Gothic War*, 4th century AD).

Essentially, Thule was an island in the North Atlantic, some six-days sail from the Orkney Islands and was ten times the size of Great Britain. The early historians declared that there were giant forests there, many wild animals and several races of men, some of whom were very primitive while others were more civilized, though they practiced human sacrifice. It was also noted that the sun did not shine for 40 days and nights in the winter, and did not set for a similar time during the summer. Curiously, a huge forest of trees, now petrified, exists on Baffin Island, further confusing any ideas about the last ice age, where, strangely, huge herds of woolly mammoths and rhinos were wandering central Alaska and huge forests existed on Baffin Island in the arctic, yet giant glacial ice sheets were in Michigan and Wisconsin. And now, even more strangely, it appears that Greenland may have been free of ice at the same time!

The lost land of Thule was important in Norse mythology, to the Teutonic Knights of the Middle Ages, and eventually to inner occult groups in Nazi Germany. The Thule Society, of which Adolf Hitler and other high ranking Nazis were members, met in Berlin and formed an occult core to the Nazi movement.

It was believed that Thule (generally identified as Greenland) had been an island to the north of Atlantis, to which many Atlanteans had fled just prior to the destruction of their land. Did the Zeno brothers map come from such a time, after the sinking of Atlantis, but before the ice had completely covered Greenland, the world's largest island? Where would these Italian navigators have run across such a thing?

As we have noted, Antonio Zeno was in the service of Henry Sinclair. Perhaps Sinclair shared with him the chart that became the brothers' blueprint. But how did a highly accurate map of the North Atlantic get to Rosslyn of the Sinclairs? It is probable that it arrived with the refugee Templars.

§§§

The Templar Connection

The reason for assuming this is simple and, I think, justified. Two mappamundi turned up in Middle Eastern archives relatively recently: the Hadji Ahmed Map in 1860 and the Piri Re'is Map in 1929. There must have been many more such maps in those same Middle Eastern archives about 900 years ago when the Templars captured and sacked many Saracen towns and cities. It is a virtual certainty that at least some similar maps were discovered by Templar founder de Bouillon's dynasty. De Bouillon was a descendant of the Merovingian kings, and therefore his dynasty and bloodline was the "Holy Grail."

The value of these ancient charts would have been appreciated at once by the Templars. Had the Kingdom of Jerusalem survived, its wealth and future prosperity would have obviously depended upon trade and commerce, not upon agriculture. Palestine in the 12th century was much as it is today. Anything that could give the de Bouillons an edge in trade would have been a treasure, and the Templars would have been given the task of guarding it.

But then the de Bouillons lost Jerusalem. They fell back on Provence to be massacred during the Albigensian Crusade. Some few representatives of this supposed Holy Bloodline allegedly survived, but then the Templars themselves were crushed and dispersed. The maps would then have had an even greater worth than they had back in Palestine. The fundamental survival of the "Holy Grail" depended upon them. If there was land across the Atlantic, as both the Hadji Ahmed and Piri Re'is Maps show, then a truly secure haven from the Inquisition potentially existed. It was the Templars' job to find that haven if at all possible, and so they fled to Scotland and Portugal with their precious maps. Probes were immediately launched out into the Atlantic.

We know that Henry Sinclair had such a map because it was included in the documented account of his voyage known as The Zeno Narrative. More on this in the next chapter.

§§§

The Mysterious Chart of Magellan

It is absolutely certain, too, that the royalty of Portugal, who were all members of the Knights of Christ which the Templars had become, had maps that showed discoveries "in advance."

Pigafetta, a navigator attached to the 1519 Magellan expedition, had this to say about a mysterious map in the possession of Magellan, copied from a map held by the King of Portugal: "The sentiments of every person in the fleet were that [the Strait of

Magellan] had no issue in the west; and nothing but the confidence they had in the superior knowledge of the commander could have induced them to prosecute the research. But this great man, as skillful as he was courteous, knew that he was to seek for a passage through an obscure strait: this strait he had seen laid down on a chart of Martin de Boheme, a most excellent cosmographer, which was in the possession of the King of Portugal."

Where had this chart come from? No one knows for certain. There is no proof, but it seems at least probable that this chart came into Portugal with refugee Templars—just as it seems equally probable that Henry Sinclair's map of the north came to the Rosslyn refuge with dispersed Templars. The coincidences are highly suggestive, even if they do not constitute absolute proof.

Columbus had a map of some sort as well, possibly even a copy from the same maps from which the Piri Re'is map was formulated. Or perhaps a map similar to that of Hadji Ahmed. Both of these show the New World, and neither confuses the Americas with Asia.

Hapgood observes that Columbus would probably not have understood the projection employed on a map like the Piri Re'is one, but that mattered little. The relative distances are accurate in azimuthal equidistant projection. Hapgood theorized that Columbus would have recognized Europe and the Mediterranean world easily enough and could have worked out a rough scale for the entire map based on the European distances he knew. He would know the direction to sail and he would know approximately the distance to his destination.

The thesis, therefore, is that these inexplicable mappamundi originated from the Templars. It seems highly probable that the Templars found such maps in the Middle East, and may have acquired some from the Nizari Assassins who were known to collect maps, books, and all forms of knowledge. Many were probably copies of earlier maps from the great libraries such as that in Alexandria.

Sailors tell tales, and maps tell tales. A black flag with a skull and crossbones flew over the Templar fleet in the Orkney Islands. It was decades after the suppression of the Templars.

Only the fleet and the castles in Scotland and Portugal were left for the Templars. The year was 1391 and Prince Henry Sinclair, Grand Master of the Templars and admiral of the Scottish fleet, met with Nicolo Zeno. Here they hatched a plan to send the fleet to North America nearly a hundred years before Columbus.

And so one of the great adventures in history was to begin: the daring exploration of unknown seas by the lost Templar Fleet; the founding of a small colony; a Vatican assassination; and the unleashing of a powerful renegade navy against the enemies of the Templars.

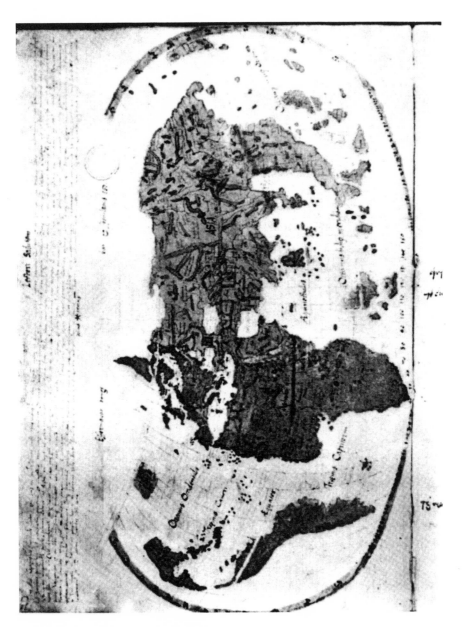

Glareanus World Map of 1510.

Martin Waldesseüller's map of 1507 on which the name "America" appears for the first time.

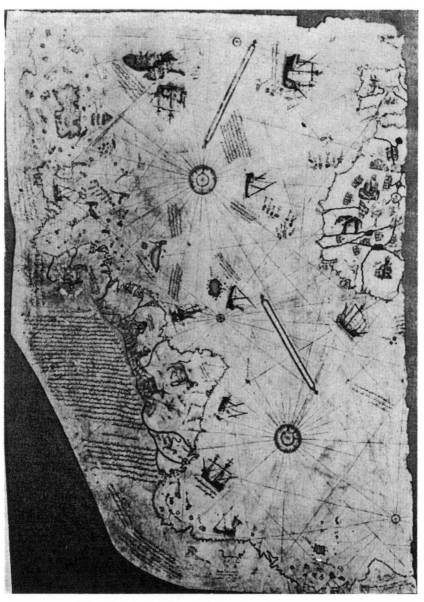

The Piri Re'is Map Number One.

The Piri Re'is Map Number Two.

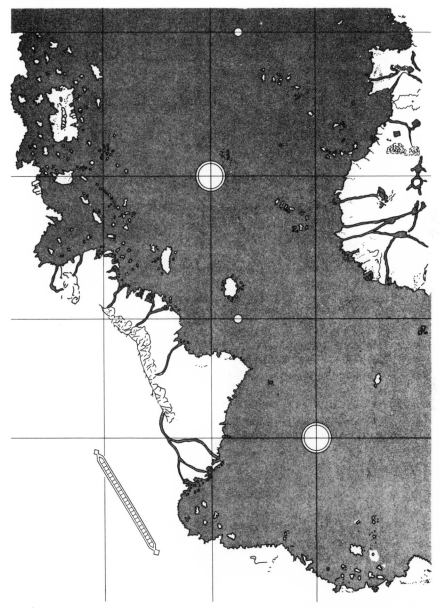

The Piri Re'is Map with its Main Grid of the Portolan Design, traced by Hapgood.

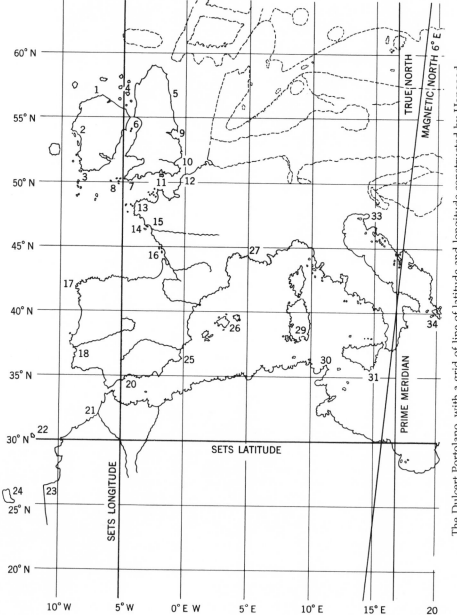

The Dulcert Portolano, with a grid of line of latitude and longitude constructed by Hapgood.

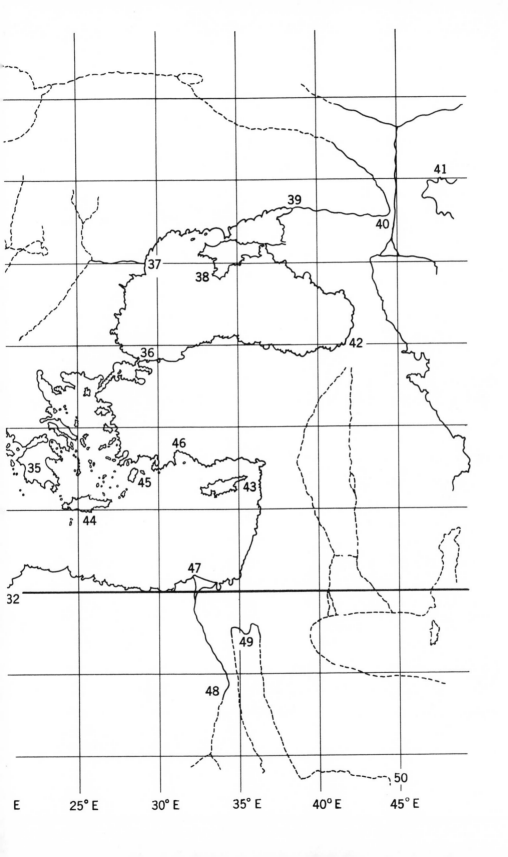

41

39

40

37

38

36

42

46

35

45

43

44

47

49

48

50

E   25° E   30° E   35° E   40° E   45° E

The World Map of Ptolemy

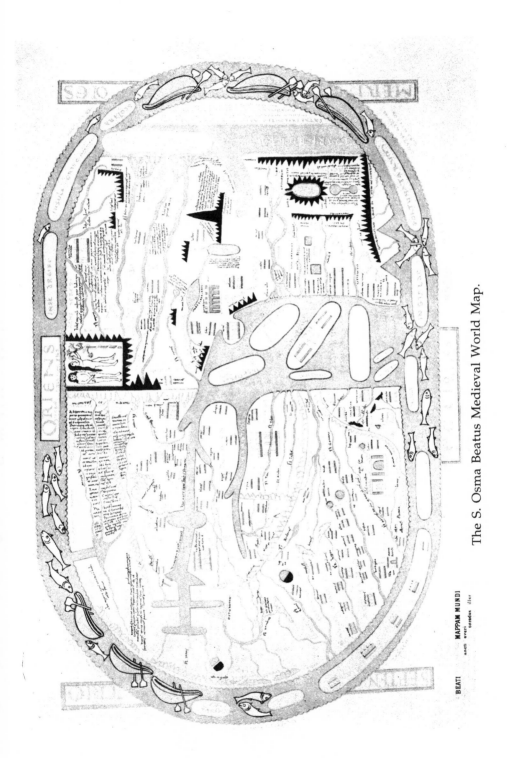

The S. Osma Beatus Medieval World Map.

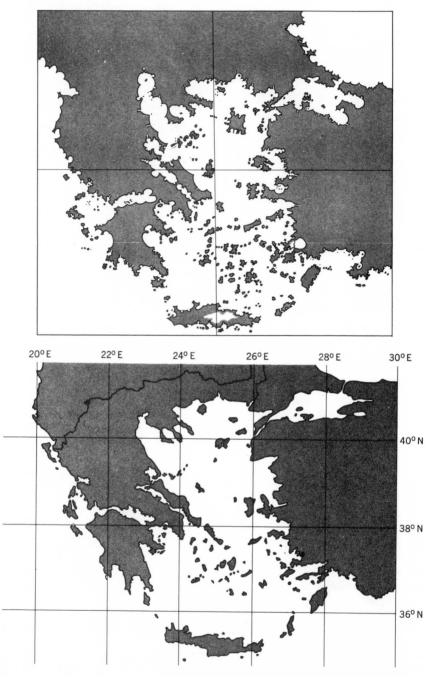

20° E	22° E	24° E	26° E	28° E	30° E

40° N

38° N

36° N

Hapgood's comparison of the Ibn Ben Zara Map, Aegean section (top) with a modern map.

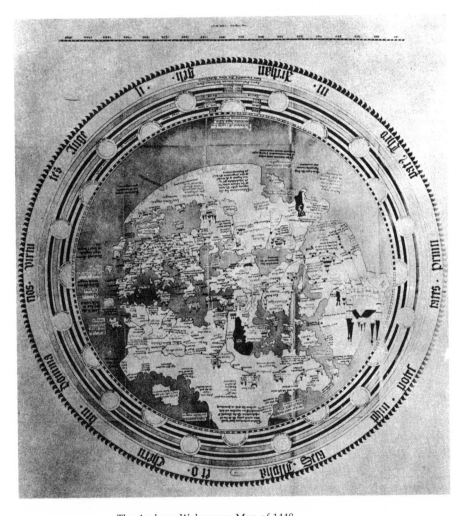

The Andreas Walsperger Map of 1448.

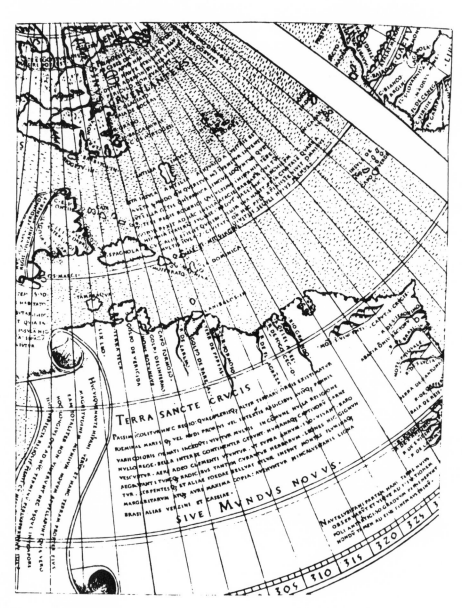

The Joannes Ruysch Map of America of 1508.

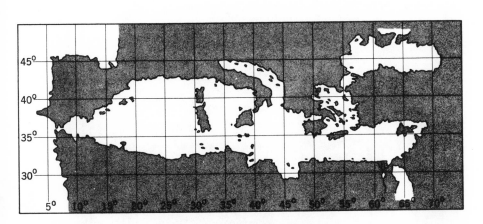

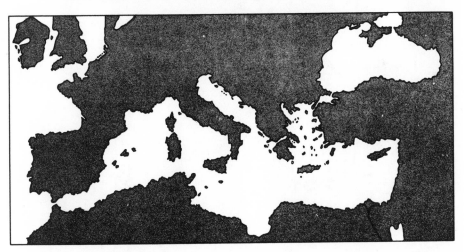

Nordenskøld's comparison of Ptolemy's Map of the Mediterranean (top) with the Dulcert Portolano.

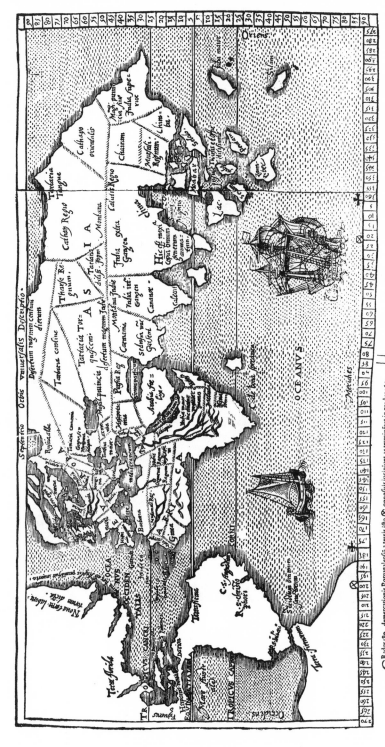

The Robert Thorne Map of 1527.

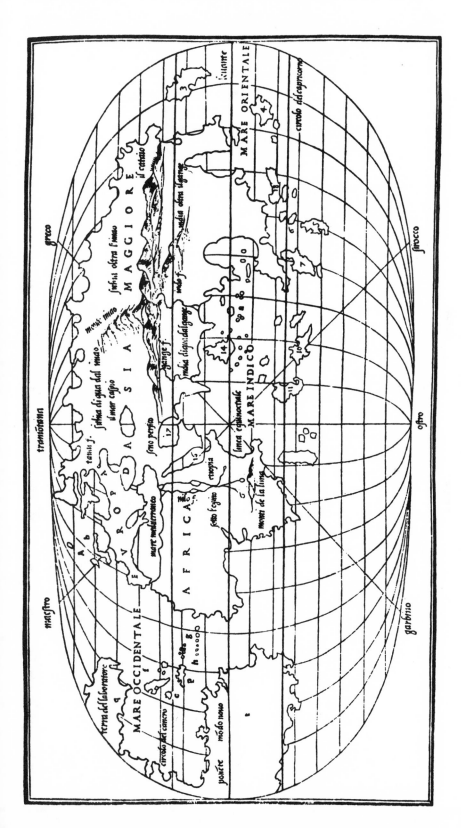

The Benedetto Bordone Map of 1528.

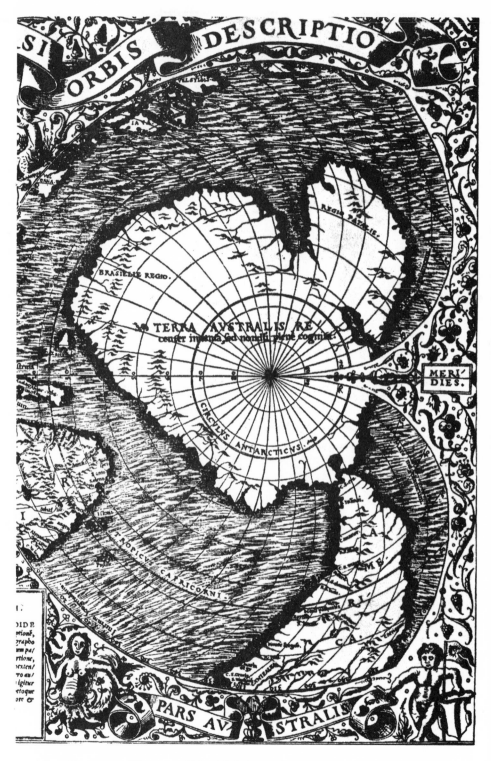

The Oronteus Finaeus World Map of 1532 showing Antarctica.

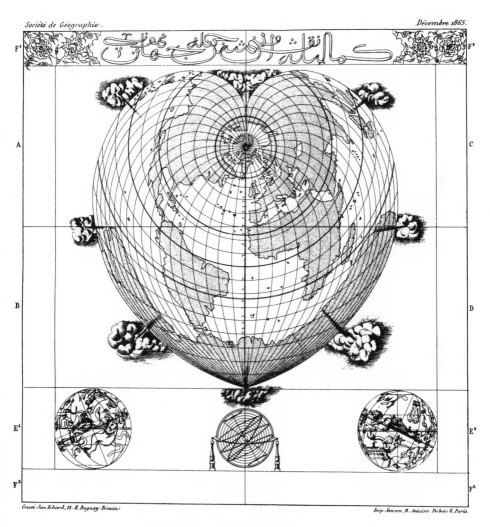

Gravé chez Erhard, 12. R. Duguay-Trouin. Imp. Janson, R. Antoine Dubois 6, Paris.

The Hadji Ahmed World Map of 1559.

The Zeno Map of the North (1380 AD).

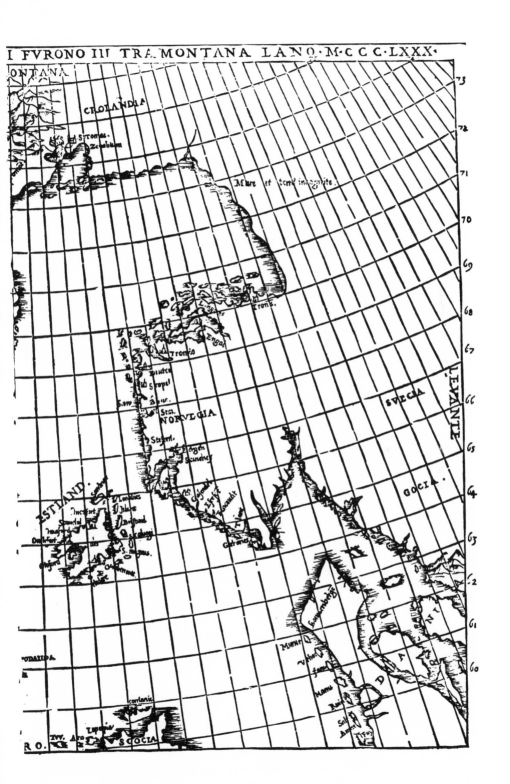

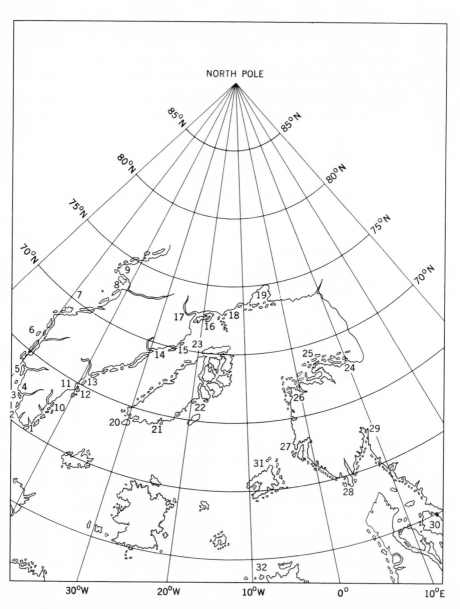

The Zeno Map with a reconstructed polar projection by Charles Hapgood.

6.
Pirates, Templars and the New World

Money is the most egalitarian force in society.
It confers power on whoever holds it.
—Roger Starr

We march backwards into the future.
—Marshall McLuhan

The Origin of the Sinclair Family

The role of the Templar fleet in the settlement of North America cannot be underestimated. And the family that set this great fleet into motion, and later founded the Masonic Lodges that would spread to New Scotland and New England, was the Sinclair family.

The Sinclairs of Scotland descend from Rognvald "The Mighty," Earl of Mocre in Norway and of the Orkneys. In 876 his son Hrolf "The Ganger" entered the River Seine and pillaged the French countryside. King Charles "The Simple" of France made peace by granting him the Province of Normandy, and the overlordship of Brittany. The treaty was signed at Castle St. Clair-Sur-Epte in Brittany from which place the St. Clairs (Sinclairs) take their name.

William the Conqueror was a first cousin of the St. Clairs. Nine Sinclair knights fought with him at the Battle of Hastings. The St. Clairs were granted vast estates in England, and their presence in Scotland pre-dates the Norman Conquest. William St. Clair accompanied Edgar "the Atheling" to Hungary and later escorted Edgar's sister Margaret to Scotland with the Holy Rood, part of the true cross. William was granted the land of Rosslyn "in liferent" by the King in 1057. Rosslyn has been in the family's hands ever since, and is the site of Rosslyn Chapel, which figures heavily in the Templar tradition.

The Chapel was built between 1446 and 1450, and is widely regarded as the crowning jewel of Freemasonry—it is full of Templar imagery, and it is said that its very dimensions utilize ancient sacred geometry. The Chapel has sparked endless debate over which Templar secrets are embedded in its stone.[12, 13, 55, 79]

William was succeeded by Henri de St. Clair, who took part in the first crusades and the Fall of Jerusalem in 1096. This was the beginning of the Sinclair involvement with the Knights Templar. The Knights were welcomed to Scotland by Robert the Bruce after the Order had been suppressed by the Pope in 1307. The Sinclairs allowed them to set up a headquarters at Ballintradoch, on the Rosslyn Estate.

At the battle of Bannockburn in 1314, recounted in chapter 3, Sir William Sinclair and two sons led out the Knights Templar to rout the English Army. The victory is still remembered by the modern-day orders of the Knights Templar at a ceremony at Bannockburn on St. John's day. Upon Robert the Bruce's death, two Sinclairs were entrusted with the task of taking his heart to the Holy Land for burial. During the journey through Spain with Sir James Douglas and a company of knights, they fought a fierce battle against the Moors. Both James Douglas and William Sinclair were killed. The Moors, impressed by the courage of the Scottish knights, allowed the dead to be returned to Scotland for burial.

The Sinclairs have a long and distinguished history in Scotland, where their family tree was spreading. In 1455 the Earldom of Caithness was granted to William St. Clair of Orkney, although there is little doubt that there were Sinclairs in Caithness long before this date. The Sinclairs of Dun are said to have settled there in 1379. Through the years, the family produced many notable persons, amongst whom was Sir John Sinclair of Ulbster, first President of the Board of Agriculture in the 1700s and the compiler and editor of the First Statistical Account of Scotland.

The Sinclairs also took an active role in the settlement of the New World and the determination of new nations. One famous descendant was Major General Arthur St. Clair, who served with Amherst at Louisberg and with Wolfe at Quebec as the British fought the French for control of territory in what is now Canada. During the American War of Independence he was a trusted adviser of General Washington and served at many battles. This is not surprising given that the Sinclair name was synonymous with Scottish Freemasonry and most of the American founding fathers were avid Freemasons.

Incredibly, Maj. Gen. Arthur Sinclair was one of the early Presidents of the United States! Because there was an eight year gap between the founding of our nation and the ratification of the constitution (1781-1788) there were 8 people who held the position of elected president before George Washington. In each of those years the President of the US Continental Congress was the prime authority of the nation. They were elected for one year. Those men in order were: John Hansen, Richard Henry Lee, Elias Boudinot, Thomas Mifflin, John Hancock, Nathaniel Gorhan, Arthur Sinclair, and finally Cyrus Griffin. Then in 1789 George Washington became President, essentially the ninth President of the United States. General Arthur St. Clair (Sinclair) was the seventh President of the United States, and the hereditary heir of the Knights Templar. Later he was the Governor of the Northwestern Territory of the United States.

It is generally thought by most historians that the esoteric cults of Freemasonry, Rosicrucianism and the Illuminati of Bavaria are actually neo-Templar organizations that were formed to preserve and popularize secret Templar knowledge. Given what we have seen of the Sinclair history, it seems to be no coincidence that the St. Clairs of Rosslyn, long-time paladins of the dynasty of Godfroi de Bouillon and the Templars, also became the hereditary leaders of Scottish Freemasonry. It is also no coincidence that Scottish Freemasonry took such a strong hold in Canada and the New World.

§§§

Prince Henry Sinclair's Secret Voyage

One of the most interesting characters in the long line of Sinclairs was Prince Henry Sinclair, the last king of the Orkney Islands. Henry, like many other nobles of the Middle Ages, held many titles and he was many things. He was "king" of the Orkney Islands, although they were officially an earldom granted to him by the King of Norway. At the same time Prince Henry held other territories as a vassal of the Scottish king. Prince Henry Sinclair was also a Grand Master of the Knights Templar, a veteran of the crusades and, according to some sources, the possessor of the Holy Grail.[12]

The Sinclair family was still strongly tied to the French families that had formerly been Knights Templars, especially with the House of Guise-Lorraine. The suppression of the Templars and the siezing of their property was still taking place in France. The Sinclairs were fighting for the now disbanded organization and were resisting the Pope's efforts to have all Templar property in Scotland seized and granted to the rival order of the Hospitallers (Knights of St. John, later known as the Knights of Malta). The Sinclairs claimed that the Templars owned no property in Scotland, though they admitted that the Sinclair clan itself did own more than 500 properties in the country, including a number of castles. One such castle was Kirkwall in the Orkneys.

As time wore on, things became increasingly difficult. The Templar groups, now widely separated, found themselves more and more isolated from the original French families that had founded the Templar Order. The Sinclairs, now the leaders of the Templars and perhaps harboring their famous treasure, were becoming more and more isolated from France, partly because a hostile England lay between them and the ports of France, and partly because the French ports were under the control of the Vatican.

In the year 1391, Prince Henry Sinclair met with the famous explorer and mapmaker Nicolo Zeno at Fer Island, a lonely spot between the Orkneys and the Shetlands. As we have seen, Zeno and his brother were well known for their maps of Iceland and the Arctic. Prince Henry contracted with them to send an exploratory fleet to the New World.[12]

With the aid of the other Knights Templar, Prince Henry gathered a fleet of twelve ships, fitted with cannon, for the voyage. The party was led by Prince Henry under the guidance of Antonio Zeno.

§§§

The Lost Treasure of the Knights Templar

So why was Sir Henry, Grand Master of the Knights Templar, so keen on sending the lost Templar fleet to North America? What did he intend to do there? Was it to found a new colony where the Templars could live in peace? Did he intend to take the Templar treasure to safety in a faraway land that the Vatican could not reach?

In his book *Holy Grail Across the Atlantic,*[9] Michael Bradley attempts to show that a vast treasure was transported to the new world by Henry. As we have seen, he claims that the treasure had been kept at Montségur in the French Pyrenees. This mountain fortress was besieged by the forces of Simon de Montfort and the Inquisition on March 16, 1244, but Bradley believes that the secret treasure escaped harm, having been spirited off by the Knights Templar. As we have noted, the treasure may have included both ancient treasure from the Middle East and the bloodline of Jesus.

While some claim that the German S.S. retrieved sacred relics from Montségur (such as the Holy Grail) during World War II, it does seem more likely that the most important parts of the physical treasure, like the Chalice Grail and the Ark of the Covenant, had already been rescued by the Merovingian kings during or after the 1244 siege. Treasure left by other groups such as the Cathars in the vicinity of Montségur may well have been discovered by the Nazis.

Bradley maintains that the early transatlantic expeditions were an attempt to find a new land, even start a new country, where the Templars and the Holy Blood-Merovingian royal family could live without persecution. Says Bradley, "Templars had been created by the second King of Jerusalem, Baudoin, the younger brother of Godfroi de Bouillon. The whole mission of the Knights Templar was to protect the dynasty of de Bouillon. That was their entire reason for being. This dynasty had been all but destroyed during the Albigensian Crusade, and the position of any survivors became more precarious with the crushing, dispersal and disbanding of their Templar guardians. But the Templars were nothing if not utterly loyal and courageous. They had always been hand-picked for just those qualities.

"If transatlantic expeditions set out from Portugal and Scotland shortly after refugee Templars arrived in both places, it is reasonable to speculate that Templars were attempting to find a haven for the survivors of the de Bouillon dynasty that these knights were sworn to protect. Survivors evacuated from Montségur, if there were any as legend maintains, desperately needed a secure refuge from the Inquisition. No place in Europe could really be considered safe in the long term. All of Europe was dominated by the Roman Church. Fringe areas, like Portugal and Scotland, might be 'safe' in the short term but only as temporary springboards to some more secure haven. The south-eastern part of Europe and the Mediterranean was no haven because the Saracens controlled it. Central and northeastern Europe were out because these regions were fighting for their very existence against the Mongols, Tartars and Turks. There was only one direction left: west. Westward, across the Atlantic, there might be a land free of the Roman Church, where a truly secure haven could be established."[9]

§§§

The Zeno Narrative and the Lost Templar Fleet

Nicolo and Antonio Zeno corresponded regularly with their brother Carlo in Venice. This correspondence became known to historians as the "Zeno Narrative." According to the Zeno Narrative, as far back as 1371, four fishing boats owned by Sinclair subjects were blown out to sea and came ashore on a distant land far to the west (probably Newfoundland). After spending twenty years there and on lands to the south, one of them was picked up by European fishermen and returned to Scotland. Note that the year would have been 1391, they year Sinclair met with Nicolo Zeno to arrange a transatlantic excursion. Even if he hadn't had prior knowledge of lands across the ocean, the return of the fishermen would have provided it with certainty. Sinclair resolved to explore these lands, and set sail with a fleet of twelve vessels, thought to be part of the lost Templar fleet.

Part of the Zeno Narrative, by Antonio Zeno, tells of their expedition:

> The nobleman [Sinclair] is therefore resolved to send forth a fleet toward those parts, and there are so many who desire to join in the expedition on account of the novelty and strangeness of the thing that I think we shall be very strongly appointed without any public expense at all.

> I set sail with a considerable number of vessels and men, but had not the chief command, as I had expected to have, for Sinclair went in his own person.

> Our great preparations for the voyage to Estotiland were in an unlucky hour; for exactly three days before our departure, the fisherman died who was to have been our guide: nevertheless, Zichmni [Sinclair] would not give up the enterprise, but in lieu of the deceased fisherman, took some sailors who had come out with him from the island.

> Steering westward, we sighted some islands subject to Frislanda, and passing certain shoals, came to Ledovo, where we stayed seven days to refresh ourselves and furnish the fleet with necessities. Departing thence, we arrived on the first of April at the island of Ilofe; and as the wind was full in our favor, pushed on. But not long thereafter, when on the open ocean, there arose so great a storm that for eight days we were continuously in toil, and driven we knew not where, and a considerable number of vessels were lost to each other. At length, when the storm abated, we gathered together the scattered vessels, and sailing with a prosperous wind, we sighted land

on the west.

Steering straight for it, we reached a quiet and safe harbor, in which we saw a very large number of armed people, who came running, prepared to defend the island.

Sinclair now caused his men to make signs of peace to them, and they sent ten men to us who could speak ten languages, but we could understand none of them, except one who was from Iceland.

Being brought before our Prince and asked what was the name of the island, and what people inhabited it, and who was governor, he answered that the island was called Icaria, and that all the kings there were called Icari, after the first king, who was the son of Daedalus, King of Scotland.

Daedalus conquered the island, left his son there for king, and gave them those laws that they retain to the present time. After that, when going to sail further, he was drowned in a great tempest; and in memory of his death that sea was called to this day the Icarian Sea, and the kings of this island were called Icari. They were content with the state which God had given them, and would neither alter their laws nor admit any stranger.

They therefore requested our Prince not to attempt to interfere with their laws, which they received from that king of worthy memory, and observed up to the present time; that the attempt would lead to his own destruction, for they were all prepared to die rather than relax in any way the use of those laws. Nevertheless, that we might not think that they altogether refused intercourse with other men, they ended by saying that they would willingly receive one of our people, and give him an honorable position among them, if only for the sake of learning our language and gaining information as to our customs, in the same way as they had already received those ten other persons from ten different countries, who had come into their island.

To all this our Prince made no reply, beyond inquiring where there was a good harbor, and making signs that he intended to depart.

Accordingly, sailing around about the island, he put in with all his fleet in full sail, into a harbor which he found on the eastern side. The sailors went ashore to take in wood and water, which they did as quickly as they could, for fear that they might be attacked by the islanders and not without reason, for the inhabitants made signals to their neighbors by fire and smoke, and taking their arms, the others coming to their aid, they all came running down to the seaside upon our men with bows and arrows, so that many were slain and several wounded. Although we made signs of peace to them, it was of no use, for their rage increased more and more, as though they were fighting for their own existence.

Being compelled to depart, we sailed along in a great circuit about the island, being always followed on the hilltops and along the seacoasts by a great number of armed men. At length, doubling the north cape of the island, we came upon many shoals, amongst which we were for ten days in continuous danger of losing our fleet, but fortunately all that time the weather was very fine. All the way till we came to the east cape we saw the inhabitants still on hilltops and by the sea coast, howling and shooting at us from a distance to show their animosity towards us.

We therefore resolved to put into some safe harbor, and see if we might once again speak with the Icelander; but we failed in our object; for the people more like beasts than men, stood constantly prepared to beat us back if we should attempt to come on land. Wherefore, Sinclair, seeing that he could do nothing, and if he were to persevere in this attempt, the fleet would fall short of provisions, took his departure with a fair wind and sailed six days to the westwards; but the winds shifting to the southwest, and the sea becoming rough, we sailed four days with the wind aft, and finally sighted land.

As the sea ran high and we did not know what country it was, we were afraid at first to approach it, but by God's blessing the wind lulled, and then there came on a great calm. Some of the crew pulled ashore and soon returned with great joy with news that they found an excellent country and a still better harbor. We brought our barks and our boats to land, and on entering an excellent harbor, we saw in the distance a great hill that poured forth smoke, which gave us hope that we should find some inhabitants in the island. Neither would Sinclair rest, though it was a great way off, without sending one hundred soldiers to explore the country, and bring us an account of what sort of people the inhabitants were.

Meanwhile, we took in a store of wood and water, and caught a considerable quantity of fish and sea fowl. We also found such an abundance of bird's eggs that our men, who were half famished, ate of them to repletion.

While we were at anchor there, the month of June came in, and the air in the island was mild and pleasant beyond description; but as we saw nobody, we began to suspect that this pleasant place was uninhabited. To the harbor we gave the name Trin, and the headland which stretched out into the sea was called Cape Trin.

After eight days the one hundred soldiers returned, and brought word that they had been through the island and up to the hill, and that the smoke was a natural thing proceeding from a great fire in the bottom of the hill, and that there was a spring from which issued a certain substance like pitch, which ran into the sea, and that thereabouts dwelt a great many people half wild, and living in caves. They were of small stature and very timid. They reported also there was a large river, and a very good and safe harbor.

When Sinclair heard this, and noticed the wholesome and pure atmosphere, fertile soil, good rivers, and so many other conveniences, he conceived the idea of founding a settlement. But his people, fatigued, began to murmur, and say they wished to return to their homes for winter was not far off, and if they allowed it once to set in, they would not be able to get away before the following summer. He therefore retained only boats propelled by oars, and such of his people as were willing to stay, and sent the rest away in ships, appointing me, against my will, to be their captain.

Having no choice, therefore, I departed and sailed twenty days to the eastwards without sight of any land; then, turning my course towards the southeast, in five days I sighted on land, and found myself on the island of Neome and knowing the country, I perceived I was past Iceland; and as the inhabitants were subject to Sinclair, I took in fresh stores and sailed in three days to Frislanda, where the people, who thought they had lost their Prince, in consequence of his long absence on the voyage we had made, received us with a heavy welcome.... Concerning those things that you desire to know of me, as to the people and their habits, the animals, and the countries adjoining, I have written about it all in a separate book, which please God, I shall bring with me. In it I have described the country, the monstrous fishes, the customs and laws of Frislanda, of Iceland, of Shetland, the Kingdom of Norway, Estotiland and Drogio; and lastly, I have written... the life and exploits of Sinclair, a Prince as worthy of immortal memory as any that ever lived, for his great bravery and remarkable goodness.[11, 12]

Although the Zeno Narrative gives no date of voyage, it is known that it must have taken place sometime around the end of the 14th century. Historian and author Frederick Pohl found a way of dating the expedition. Noting that it was common for explorers to name their discoveries from the religious calendar, Pohl seized on the statements "while we were at anchor there, the month of June came in" and "To the harbor we gave the name Trin." Might "Trin" refer to Trinity Sunday which is the eighth Sunday following Easter? Pohl figured that since historical records indicate that Henry Sinclair died in August 1400, his voyage was earlier. Pohl searched the dates of past Easter celebrations and selected June 2, 1398, from the following Trinity Sundays: June 6,

1395; June 28, 1396; June 17, 1397; June 2, 1398; May 25, 1399; June 13, 1400.[11]

Pohl figured that June 2, 1398, was the most likely date because it is the closest day to when "the month of June came in."

Canadian historian William S. Crooker thinks that it may have been several years earlier as he explains in his book *Oak Island Gold:* "For my own part, I can't be sure. Pohl may have thought that two years was long enough to be away from home but I think it is relevant that Antonio says on returning to Frislanda (the Orkney Islands of Scotland), 'the people, who thought they had lost their Prince, in consequence of his long absence on the voyage.' Would two years have been considered a 'long absence'? Could the date have been June 6, 1395? The month of June might have been considered to have 'come in' while at anchor on the first week, and Trinity Sunday would have still been an appropriate day from which to take a name. However, Pohl's method of deduction is clever and most historians accept June 2, 1398. But my comment should be taken seriously by those who feel that Sinclair may have needed more time than two years to accomplish all for which he is credited."[14]

§§§

Prince Henry in "New Scotland"—Nova Scotia

Nova Scotia is thought to be the second "island" visited on Henry Sinclair's expedition because of the Zeno Narrative statements, "we saw in the distance a great hill that poured forth smoke" and "the smoke was a natural thing proceeding from a great fire in the bottom of the hill, and that there was a spring from which issued a certain substance like pitch, which ran into the sea, and that thereabouts dwelt a great many people half wild, and living in caves." In 1951, Dr. William H. Hobbs, a University of Michigan geologist, pointed out that the only pitch deposits in the coastal region of North America were at Stellarton and Pictou in Nova Scotia.[11, 14]

Pitch deposits of Stellarton do indeed run into the tidal river, and in the past, the Stellarton pitch deposits have burned out of control. And, it was only in the Stellarton region that the Micmac natives lived in caves.

Presuming that the "great hill that poured forth smoke" is Mount Adams, Pohl drew a straight line from the Stellarton pitch deposits through the mountain to the Atlantic Ocean. His line struck Chedabucto Bay and he concluded that the harbor "Trin" where Henry Sinclair landed was most likely Guysborough Harbour at the head of the bay.

The Zeno Narrative tells nothing of Henry Sinclair's activities after landing in Nova Scotia. Antonio Zeno writes only of his own return to Frislanda. However, the narrative indicates that Sinclair was interested in exploring the country and making contact with its inhabitants. He sent one hundred of his men to investigate the "great hill that poured forth smoke," to explore the country and find out "what sort of people the inhabitants were." His men returned with word of people living in caves and a "very good and safe harbor." Believing the land he had found was an island, it seems reasonable that he followed the shore around to the Stellarton area where he ingratiated himself to the Indians and used their help to explore the new country.

Pohl constructed Sinclair's comings and goings from Micmac legends about the man-god, Glooscap. Pohl identified Glooscap as Henry Sinclair from a long list of specific similarities, one being that they both had three daughters. Other scholars had accepted Glooscap as being a European, so Pohl's identification is not much of a stretch.

In his book, *Prince Henry Sinclair*,[11] Pohl attempts to trace Glooscap's movements. Micmac legends say that before the onset of winter, Glooscap made a return trip across the Nova Scotia peninsula from the Bay of Fundy to the Atlantic. Pohl surmises that Glooscap (Henry Sinclair) paddled west along the Bay of Fundy shore of Nova Scotia to the Annapolis Basin and the present site of Digby. From Digby, he paddled east to Annapolis Royal. From Annapolis he reached the Atlantic via the lake system

that crosses the peninsula to the Mersey River, which empties into the ocean at Liverpool, Nova Scotia. This is a logical route for it provides an almost uninterrupted waterway. One of the lakes in the system is Lake Rossignol, the largest lake on the Nova Scotia peninsula. Following the crosscountry exploration, Sinclair wintered on Cape d'Or on the Bay of Fundy.

Michael Bradley, in his book *Holy Grail Across the Atlantic*,[9] suggests that Pohl is wrong about the return route, suggesting that a good explorer would probably take a different route back so as to explore as much of the country as possible. He therefore takes the liberty of revising Pohl's map of Sinclair's journeys to show the shorter return crossing by the Gold and Gaspereau Rivers. This would bring Sinclair's party into the New Ross area where Sinclair supposedly built a mini-castle.

Bradley asserts that Prince Henry took over as many as 300 colonists to the New World and built a literal "Grail Castle" in Nova Scotia. So strong is the evidence for Prince Henry Sinclair's voyage across the Atlantic with the Knights Templar that his distant relative Andrew Sinclair wrote a book entitled *The Sword and the Grail*[12] in which he agreed with Bradley.

Bradley and Sinclair claim that the special Grail Castle was built in an area of central Nova Scotia called "The Cross." This spot could be reached via river from either side of the Nova Scotia peninsula and at the mouths of both rivers was an island called "Oak Island." Curiously, one of these Oak Islands has the famous "Money Pit" which is a man-made shaft hundreds of feet deep with side tunnels. It is believed that there is a treasure in this pit and millions of dollars have been spent in attempts to reach the submerged bottom. More on that later.[12]

In 1974 the apparent remains of a 14th-century castle were discovered in mid-peninsular Nova Scotia and reported to the provincial government. In 1981, Bradley was asked to look at the site by Nova Scotia's Ministry of Culture, and launched an investigation that lasted until April 1983.

Bradley continues to promote the idea that the remains of a pre-Columbian European fortification exist on a hilltop location in Nova Scotia. He says that the ruins appear to be rubblework construction appropriate to Sinclair's late 14th-century era, and typical of other Northern Scottish and Scandinavian fortifications of the period; and that several other sites were also discovered which seem to have been lookouts for the main fortress or settlement.

Says Bradley, "The two-and-a-half-year investigation recommended excavation of the site by archeologists from the Nova Scotia Museum, but this has not yet been done. I hasten to point out that the other sites seemingly associated with the major mid-peninsular construction were not discovered by me and my associates, but by Frederick Pohl and a team from the University of Maine in 1959-1960. There is also evidence, again mostly uncovered if not actually discovered, by Frederick Pohl, which indicates that the Sinclair expedition at least casually explored as far south as Massachusetts."[9]

According to the Sinclair family, after overwintering with the Micmac Indians, Henry explored the eastern seaboard of the United States. His ships sailed to Massachusetts, where one of his knights died. An effigy of the Knight, now thought to be Sir James Gunn of Clyth, was chiselled into a rock face at Westford, Massachusetts. Henry sailed on to Rhode Island where he is thought to have built the Newport Tower.[12]

Further proof of Henry's transatlantic voyage is visible in Rosslyn Chapel where there are carvings of Indian maize and North American aloe cacti, all carved years before Columbus set sail.

§§§

The Assassination of Henry Sinclair

Prince Henry and his fleet returned to the Orkneys in the year 1400. He was in control of one of the largest and most experienced fleet of ships in the world at the time, and he was apparently using it to build a colony in North America with the intention of moving much of the Templar treasure there.

However, upon arriving back in Kirkwall castle in Orkney, Sir Henry Sinclair was suddenly killed. The method of his death is not certain. Some have suggested that he was assassinated by a special team sent to the Orkneys; other historians maintain that he died while meeting an attacking army of English who had just landed at Scapa Flow to ransack the village that was near Kirkwall castle.

Records are virtually non-existent, but it appears that in August of 1400, King Henry IV of England invaded Scotland during a period when the ever-warring clans of Scotland were out of control, with the Donald clan fighting others including the Sinclairs and the McKays clan.

One version has Sir Henry Sinclair being awakened early in the morning with news that a hostile English force had landed near Kirkwall. Puting on their armor and saddling the horses, Sir Henry and his knights imprudently charged out of their protective castle and attacked the English. Allegedly overwhelmed by the English force, many of the Sinclair Scottish force were killed, including Sir Henry.[44]

Because of sketchy records (one English document maintains that Sir Henry actually died in 1404), and a lack of evidence that a large invasion of the Orkneys had really taken place in 1400, it has been suggested that Sir Henry was killed by a special naval team sent to assassinate him, perhaps one sponsored by the Vatican.

It seems likely to me that a Vatican assassination team, complete with a commando crew of raiders, landed at Scapa Flow at this time period in order to lure Sir Henry into their trap. Perhaps only one ship full of men was dispatched, and Sir Henry, thinking that it was a normal pirate attack, felt that he could deal with the intruders. Both Frederick Pohl and Andrew Sinclair in their books describe Sir Henry as putting on his armor in the safety of Kirkwall castle and foolishly going out to meet these wayward seafarers with a few of his guard.

This may have been the case. Sir Henry may have believed that by telling the ship and its pillaging crew that they were dealing with none other than the Earl of Orkney himself they would probably just leave, whether Viking, English or Scottish raider. Instead, Sir Henry was jumped by the entire contingent of men from the ship who stabbed him numerous times until he was dead.

What is curious here is that these raiders seemed intent on murdering Sir Henry. When they were informed that he was the

Earl of the Orkneys, rather than being deterred from their criminal ways, they instead pounced upon him with great fervor. This puzzling and important incident will remain a pivot point in history concerning naval warfare, control of the New World, and the rise of piracy.

If Sir Henry had lived, he might have gone on to declare to the world that there was another land across the North Atlantic. Author Steven Sora in *The Lost Treasure of the Knights Templar*[44] maintains, however, that Sir Henry and the Sinclair family would probably have kept the location of their new colony a secret for some time to come.

§§§

The Mystery of New Scotland and New Atlantis

Both Bradley and Sinclair claim that Canada was settled as a direct result of the Holy Grail being taken there. Sinclair and the Templars were attempting to create the prophesied "New Jerusalem" in the New World.

The French explorer and founder of the Quebec colony Samuel de Champlain (1567-1635) was a secret agent for the Grail Dynasty, says Bradley, and the Grail was moved to Montreal just before Nova Scotia was attacked by the British Admiral Sedgewick in 1654. A mysterious secret society called the Compagnie du Saint-Sacrement carried the Grail to Montreal. Its whereabouts today are unknown, according to Bradley.[9]

The many indications that Montreal was purpose-built by the Templars to be their holy city is discussed in detail in Francine Bernier's book *The Templars' Legacy in Montreal, the New Jerusalem*.[33]

The fascinating concept of the Knights Templar taking the Holy Grail to the New World in order to found the New Jerusalem takes us directly into Atlantis studies. It is possible that the exploits and aspirations of Prince Henry influenced Sir Francis Bacon who, around the year 1600, published his unfinished utopian romance entitled *The New Atlantis*.

Bacon's *The New Atlantis* tells how, on a voyage from Peru to China, the narrator's ship was blown off course to an undiscovered South Sea land where a turbaned population have a perfect society, that is, a democratic state with an enlightened king, much as that to which Britain aspired.

The people had come to their country which they called "Bensalem" from Plato's Atlantis, which was actually in America. Bacon's Atlantis-from-America utopia is a very scientific land with submarines, airplanes, microphones, air-conditioning and a great research foundation. The folk of Bensalem are Christians, of course, having received the gospel from Saint Bartholomew.

Bacon's *The New Atlantis* was published 200 years after Prince Henry's tragic murder in the Orkneys in 1400 AD. Was he drawing on rumors of a new utopia having been established in America by Henry? It certainly wouldn't be the tale told out of New Scotland.

Nova Scotia was originally named *Acadia* by the explorer Giovanni da Verrazzano in 1524. He was better educated than most of the early explorers, so it is reasonable to assume that he gave the name *Acadie* or *Acadia* to the lands he was claiming, equating them with the *Arcadia* of ancient Greece. Verrazzano's region grew to include all of Nova Scotia, southeastern Quebec and the eastern part of Maine. One theory is that Verrazzano was told by the Micmac Indians of the area that this country was called cady or quoddy, which means "land" or "territory" in the Micmac language. Hence Verrazzano used a sort of pun in the name of Acadia.

There is an interesting connection between Verrazzano's Acadia, ancient Greece and the ancient Egyptians, who were controlled by the Greeks after Alexander the Great's invasion.

According to Barry Fell in *America B.C.*,[16] the Micmacs are the descendants of Egyptian explorers and settlers who came to North America. Fell developed this theory when he was shown the Lord's Prayer written in Micmac hieroglyphics printed in 1866. The Micmac hieroglyphic system was supposedly "invented" by a French priest, but Fell recognized that the Micmac hieroglyphic system was virtually identical to the ancient Egyptian hieroglyphic system, with many, or most, of the hieroglyphs having the same meaning in both languages.

Fell devotes several pages in his book to comparing Micmac hieroglyphs to ancient Egyptian and relates that Micmac hieroglyphs were already in use in 1738 when Abbé Maillard adopted them for his *Manuel Hieroglyphique Micmac.* Fell points out that Egyptian hieroglyphs were not deciphered until 1823, when Champollion published his first paper on the Rosetta Stone. After a good deal of research, Fell was convinced that the Micmac Indians, and other Algonquian Indians had been using a system of Egyptian hieroglyphs for at least 2000 years. He quotes Father Eugene Vetromile who said in 1866, "When the French first arrived in Acadia, the Indians used to write on bark, trees, and stones, engraving signs with arrows, sharp stones, or other instruments. They were accustomed to send pieces of bark, marked with these signs, to other Indians of other tribes, and to receive back answers written in the same manner, just as we do with letters and notes. Their chiefs used to send circulars, made in the same manner, to all their men in time of war to ask their advice, and to give directions."[16]

When it comes to listing the mysteries of Nova Scotia and the Bay of Fundy, one should not ignore Henriette Mertz and her book, *The Wine Dark Sea.*[63] This scholarly and interesting book analyzes the voyage of Odysseus (Ulysses to the Romans) from Homer's ancient Greek epic *The Odyssey,* and tracks the voyage of the legendary sailor through the North Atlantic. According to Mertz's detailed itinerary, Odysseus sails out through the straits of Gibraltar and into the North Atlantic, eventually arriving at the headlands of the Bay of Fundy, which she identifies as the "monsters" Scylla and Charybdis. Homer has Odysseus attacked by Charybdis, being "sucked down the salty sea—we could see within the swirling cataclysm of the great vortex and at the bottom the earth appeared black with sand while round about the rock roared terribly..." According to Mertz, Odysseus is in reality caught in the deadly tidal bore of the Bay of Fundy.

§§§

Templars, Pirates and the Mystery of Oak Island

South of Halifax, Nova Scotia is Mahone Bay, and on an island in that bay is one of the most famous and mysterious treasures in the world. The bay is named after pirate tradition. The name

"Mahone" is derived from the Turkish word "mahone" which was "a low-lying craft, propelled by long oars, called sweepers, and much used by pirates in early days of the Mediterranean."[14]

Oak Island allegedly has a lost treasure deep within a bizarre shaft that has befuddled treasure-seekers for two centuries. In 1795, three youths rowed over from the mainland for a day of exploration. They found an oak tree with a sawed-off limb projecting over a large circular depression in the ground. They came back the next day with picks and shovels.

Their digging revealed a circular shaft about 13 feet wide. They discovered a platform of logs at the ten-foot level, and again at the 20 and 30 foot levels. As they continued the excavation over a period of many years, they discovered a platform of logs every ten feet to a depth of 80 feet. At 90 feet they discovered a round flat stone with markings that they could not decipher. The stone was later "deciphered" and allegedly read, "ten feet below two million pounds are buried." The whereabouts of this stone has not been known since 1935.

At 98 feet, confident that the treasure was near, they stopped digging for the weekend. When they returned to the shaft they were dismayed to find that the pit was now half full of seawater. They pumped out the water, reached the 110-foot level, but finding nothing, they reluctantly abandoned the island and their search.

Subsequent searches in the 1800s created side tunnels and various dams to keep the pit from flooding. Allegedly, at the 154-foot level, two chests were discovered in 1897. A depth of 170 feet was attained in 1935, and two years later, a depth of 180 feet. A second tunnel to the sea was discovered in 1942 and in 1971 a consortium out of Montreal purchased the island and reached a water-filled cavity at the 212-foot level. According to news reports of the day, a submarine television camera that was lowered into the cavity sent back images of three chests and a severed hand. Divers were lowered into the cavity but arrived too late, for in the meantime the cavity had been eroded by seawater. The quest for this treasure, one of the world's costliest, continues to this day.[14, 15]

The treasure has been variously thought to have been laid by pirates such as William Kidd, Henry Morgan or Blackbeard, or by

Norsemen. More likely, it was a British or Spanish payroll that was secreted in the shaft, waiting for the fighting to cease during one of the many wars of the Spanish Main, or even the American Revolution. If the treasure had been Blackbeard's, the secret doubtlessly went to the grave with him. In 1718 Lt. Robert Maynard of the Royal Navy was sent by Gov. Spotswood of Virginia in the sloop *Ranger* to capture Blackbeard. Maynard caught the pirate ship at Ocracoke, North Carolina and Blackbeard was killed in the battle. Perhaps, he had just finished creating the Oak Island pit.

Other more far-out theories are that the pit was made by the Incas, hiding their treasure from the conquistadors—or that it was an Atlantean tomb! In a small, privately published book, Canadian Atlantis researcher, Alexander Stang Fraser theorizes that the Grand Banks to the east of Labrador and Nova Scotia was the site of fabled Atlantis, now submerged. Fraser maintains that a flood tunnel security system, apparently designed to drown excavators at the pit site, would seem to have precluded withdrawal of the presumed treasure. "This contradiction indicates that this was not originally a well protected buried treasure situation, but a unique type of tomb. The protection of a tomb by the sea is consistent with the maritime activity of the Atlanteans, and Mahone is in the general area of the proposed location of the former Atlantis."[59]

Fraser calls the Grand Banks the "Elysian Plain" of Plato's account of Atlantis, and it is an interesting theory in that this large area, just beneath the surface of the Atlantic, was at one time above water, and may have contained a civilization. It seems unlikely, however, that the Oak Island Money Pit has anything to do with this sunken land.

In the wake of such wild theorization, it may not seem so bizarre to suggest that the Money Pit may have been dug by the Templars to secrete their treasure during the 1398 voyage to Nova Scotia. In fact, both Bradley and Sinclair make this assertion in their books.

§§§

The Black Dog with Red Eyes

William Crooker in his book *Oak Island Gold*[14] talks about a number of curious folk tales on Oak Island. Crooker says that Nova Scotia is teaming with legends of pirates, and buried treasure is deeply embedded in the collective consciousness of the people of Chester and the Mahone Bay area. An imitation treasure chest filled with costume jewelry and sundry items is displayed in a shop window; a mannequin pirate garbed in rubber boots, holding a sword and wearing the proverbial black patch over one eye, stands guard beside the entrance of a gift shop; a sign advertising Oak Island tours displays a pirate carrying a chest on his right shoulder; restaurant and beverage lounges operate under names such as "The Pirate Lure Beverage Room Grill" and "Cap'n Kidd's Lounge."

Oak Island is steeped in superstition and shrouded with stories of the supernatural based on piracy. Perhaps the spookiest story is the one about pirates who create a ghost to guard a treasure. According to legend, the pirate crew, when burying a cache, would stand around the hole into which a chest had been lowered and the captain would exclaim, "Who is going to stay and guard the treasure?" Being a greedy lot, the pirates would compete verbally for the assignment, each hoping to be chosen so that he could keep the treasure for himself once the captain and his crew departed. One voice crying, "I will" would naturally ring out louder than the rest, and the captain would identify that particular pirate, approach him and say, "You've got the job." That evening, the captain would throw a big party with loads of rum and all the crew would get roaring drunk and dance around a huge bonfire. Toward the end of the evening when most of the crew had passed out and the others were bordering on comatose, the captain would bludgeon the sozzled volunteer and push him into the pit on top of the treasure chest. The captain would then order a handful of staggering drunks, still able to function, to shovel the earth back into the hole, thus leaving a ghost behind to guard the treasure.[14]

There are various twists as to how a pirate was selected to remain as a ghost. According to another version, an unsuspecting stranger was taken ashore along with a few men and he innocently volunteered thinking that he would retrieve the ill-gotten spoils for himself as soon as the pirate ship sailed away. And, another version says that the pirates gathered together on the beach around the open pit into which the treasure chest had been lowered and drew lots to determine who should be murdered and buried with the cache. The "winner" would either be decapitated by the crew or buried alive with the treasure. An even more gruesome yarn suggests that the captain would select a man at random, cut off his head, and toss it into the pit on top of the treasure to watch it.[14]

An alternative to the ghost of a pirate guarding treasure on Oak Island is the legend that "a dog with fiery eyes" guards the treasure of Captain Kidd, which many believe lies buried on the Island. According to this story, the dog has "blood red eyes that glow like hot coals" and is thought to be the Devil's watchdog. Others believe that the dog is the ghost of a pirate who was sacrificed to the Money Pit.

Harris Joudrey, who lived his first ten years on Oak Island, claimed to have seen the dog when he was a boy of nine, in 1900. According to Joudrey, the large dog was sitting on the sill of a door to the "boiler house" of the Money Pit. The dog had never been seen before on the island. The animal watched young Joudrey and some of his companions as they walked by and disappeared from sight, and he was never seen again. The incident "scared the daylights" out of Joudrey and his friends.[14]

In another version of the legend, a lady relative of Anthony Graves, the Graves who owned most or all of Oak Island, was "scared stiff" by a dog as large as a small horse or colt. The encounter was on the north side of the island, and the relative witnessed the huge dog disappear into the side of a stone wall that marked one of the property boundary lines.[14]

We are repeatedly told that the widespread belief in ghosts guarding buried treasure made it difficult for early Oak Island expeditions to recruit workmen.

Oak Island is said to have been viewed with a superstitious eye by all of the residents of nearby Chester, and there are many curious stories related to the Island.

According to folklore, a light might indicate the place where a treasure is buried and one story tells of a bright light appearing on Oak Island and illuminating the Money Pit. Men are seen burying a treasure in the glow of the light. The phenomenon is said to be "looking back into time," as if the light were functioning as a time machine bringing the past forward for a brief observation.[14]

The belief that ghostly lights seen at night usually indicates that a treasure has been buried nearby is supported by another Nova Scotia tale. According to rumor, when a chest of doubloons (sometimes called "double loons" by early storytellers) was discovered at Port Medway, a light hung in the air above the place where it was found. After the discovery the doubloons were delivered to a bank and exchanged for a large sum of money.

That light is associated with buried treasure is further supported by another piece of folklore. Many locals believe that treasure is buried on Tancook Island in Mahone Bay, because a bright light has often been seen there before a storm.

§§§

Mysterious Ghosts in Red Jackets and the "Ancient Man"

Oak Island-inspired visions form an integral part of the local folklore. Two popular legends tell of the specters of men dressed in red jackets.

A man is said to have been treasure digging on Oak Island, but for reasons unknown he was obliged to temporarily discontinue his work. When he resumed, he was interrupted by a man in a red coat who told him that he was "not digging in the right place." The red-coated man then dematerialized by disappearing into the bottom of the Pit.

In 1940, a little girl living with her parents on Oak Island reportedly saw men at Smith's Cove wearing red jackets. She was scared and crying when she came running into the house so, her father went down to the shore to investigate. There was no one there. Furthermore, it was winter and there was snow on the ground but there wasn't a trace of a footprint anywhere.

Another vision involves what has become locally known as the "Ancient Man." Anthony Graves' son William saw this apparition in the 1850s when he was out in his boat one evening trapping lobsters on the north side of Oak Island. The Ancient Man was seated between a couple of trees near the shore. Displaying long whiskers, the apparition called to William, saying, "Come here and I will give you all the gold you can carry." William didn't take the specter up on his offer. He was so frightened that he rowed away from the island as fast as his oars would propel him. When William encountered the Ancient Man, he was in his twenties. He told the story of the Ancient Man while on his deathbed.

In another instance of Oak Island-inspired visions, a visitor to the Money Pit witnessed the murder of a Spanish monk. The victim had his throat cut and was buried in a subterranean tomb. During the vision, the tourist went into a trance, and ended up rolling around on the ground, screaming.

Franklin Roosevelt and the Oak Island Treasure

In April 1909 a man named Captain Henry L. Bowdoin formed a company to exploit the Oak Island treasure. He named the company the Old Gold Salvage and Wrecking Company and took up offices at 44 Broadway in New York. The firm's authorized capital was $250,000, divided into a public offering of shares at one dollar each.

The company's prospectus was impressive, estimating the value of the "pirate" treasure at over 10 million dollars.

Bowdoin proposed locating the supposed treasure by pinpointing the flood tunnel that allowed ocean water into the pit, and then driving interlocking sheets of piling into the tunnel to block off the water. They would do this by excavating holes into the tunnel on both the shore and inland sides of the piling, and finally pumping the inland hole dry. After blocking off the tunnel, he proposed to move his large excavation bucket to the Money Pit and, with the aid of a 1,000-gallon-per-minute pump, recover the treasure.

Conceding that this approach might fail, Bowdoin offered this alternative: "Should, however, the tunnel not be located; the pump not be able to keep out the water, and the bucket not bring up all of the treasure, some, perhaps having slipped to one side, then, and in that event one of Bowdoin's Air Lock Caissons could be placed in the pit, sunk through water or earth to any depth desired, and side tubes forced out to reach any desired spot. Compressed air keeps out all water and allows men to work at the bottom, and send any articles up and out through the air lock. The caisson is now used in sinking foundations through earth and water to bedrock; for foundations of buildings and through the water of a river and to bedrock beneath its bed; for foundations for bridges, piers, etc. The treasure can positively be recovered by its use."[14]

The prospectus ends by addressing future company plans following the work on Oak Island, and by summarizing its objectives. "The wide publicity given by the press of the intention of Mr. Bowdoin to recover this Oak Island Treasure has resulted in the receipt of a number of letters calling attention to other treasures and valuables that could be recovered through scientific modern equipment. Several of these are of exceptional interest, being valuable, non-perishable cargoes of vessels, and sunken valuables, the locations of which are known.

"As no systematic, scientific efforts have been made to recover and salvage the more or less valuable cargoes, etc., of wrecks which are piling up at the rate of over one hundred a year, it is deemed good business policy to get together a complete modern equipment, and after the recovery of the treasure at Oak Island, to utilize said equipment in a general salvage and wrecking business...

"The recovery of the [Oak Island] treasure would yield a dividend of 4,000 percent on the entire capital stock; and, as operations should begin in May or June and be completed in three or four weeks, should be available this summer. This will leave time for the salvage of an exceptionally valuable cargo before next winter, when the plant would operate in southern waters, where certain other valuables await our attention.

"To purchase the equipment, stock is now offered to the general public at a popular price, $1.00 per share. No order for less than ten shares will be accepted."[14]

Franklin Delano Roosevelt was apparently attracted by the

newspaper articles about Bowdoin and impressed with the prospectus of The Old Gold Salvage and Wrecking Company. He and three of his friends, Duncan G. Harris, Albert Gallatin, and John W. Shields, purchased shares in the company and visited Oak Island during Bowdoin's expedition. Roosevelt himself visited the Island several times in 1909.

Despite the confidence and expertise that Bowdoin radiated, he was only able to raise $5,000 through the sale of shares, but Frederick Blair helped by granting the company a two-year lease on the property in exchange for his shares. With all the confidence gained through engineering training and experience, Bowdoin was certain that he would succeed. He set sail from New York (seriously behind schedule) on August 18, 1909. After purchasing extra equipment in Halifax, they arrived at Oak Island on August 27th and set up their operation which was dubbed "Camp Kidd."

After an unsuccessful attempt to locate the flood tunnel entrance at Smith's Cove, Bowdoin made a direct assault on the Money Pit. All obstructions such as platforms and ladders were ripped out using the excavation bucket to a depth of 113 feet. Due to the lack of funds, the Company had been unable to purchase the proposed 1,000-gallon-per-minute pump. The water level in the Pit stood 30 feet below the ground surface. When a diver was lowered down to the bottom of the Pit to report on the conditions of the cribbing, the report was discouraging. The cribbing was in a terrible state. It was twisted and badly out of alignment. The bottom was covered with a mess of tangled planks; timbers were "sticking up in all directions."[14]

Bowdoin then resorted to conventional drilling. More than two dozen borings were made to depths as great as 171 feet without encountering a trace of treasure. By this time Bowdoin had exhausted all of the funds and the operation was closed down in November.

Roosevelt had probably joined the expedition more out interest, plus the adventure and romance of treasure hunting, than for the expectation of financial gain. He kept in contact with his 1909 fellow searchers for a period spanning 30 years. His personal records include letters regarding Oak Island that date as late as 1939 when he was serving his second term as president of the

Pirates & the Lost Templar Fleet

United States.

That a man of Roosevelt's intelligence and wisdom retained an interest in the Oak Island enigma for the better part of his life is a curious testament to the lure of Oak Island and its mysteries. Did Roosevelt know something about the Sinclairs and the lost Templar treasure? It was during his term as President of the United States that his Vice President, Henry A. Wallace, had the Masonic-Templar symbol of the Great Pyramid with the Eye of Horus at its peak adopted as the Great Seal of the United States.

172

English Masonic apron complete with skull and crossbones.

Masonic symbols. Note the skeleton and crossbones in foreground.

Rosslyn Castle in Scotland.

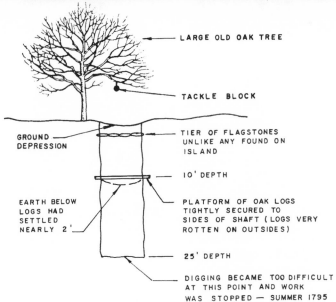

LARGE OLD OAK TREE

TACKLE BLOCK

GROUND DEPRESSION

TIER OF FLAGSTONES UNLIKE ANY FOUND ON ISLAND

10' DEPTH

EARTH BELOW LOGS HAD SETTLED NEARLY 2'

PLATFORM OF OAK LOGS TIGHTLY SECURED TO SIDES OF SHAFT (LOGS VERY ROTTEN ON OUTSIDES)

25' DEPTH

DIGGING BECAME TOO DIFFICULT AT THIS POINT AND WORK WAS STOPPED — SUMMER 1795

Top: Aerial photo of Oak Island. Bottom: The Money Pit.

Rosslyn Castle in Scotland.

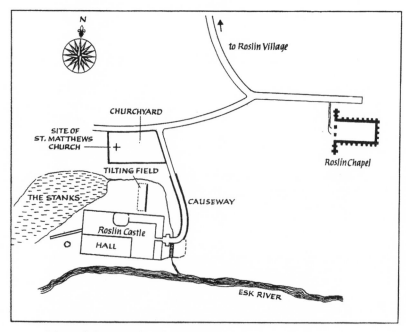

Map of the Rosslyn Castle area.

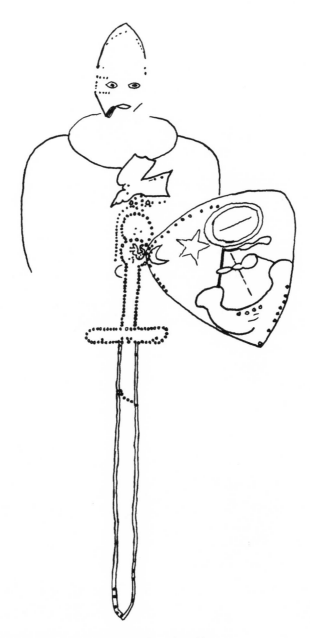

The Westford, Connecticut knight effigy.

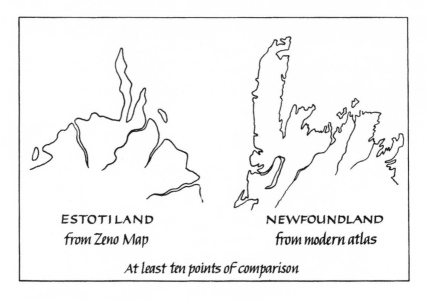

ESTOTILAND
from Zeno Map

NEWFOUNDLAND
from modern atlas

At least ten points of comparison

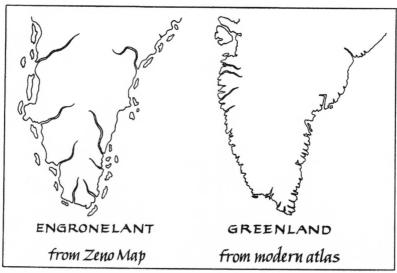

ENGRONELANT
from Zeno Map

GREENLAND
from modern atlas

Frederick Pohl's comparisons of the Zeno map and modern geography.

Maj. Gen. Arthur St. Clair of the Sinclair family.

179

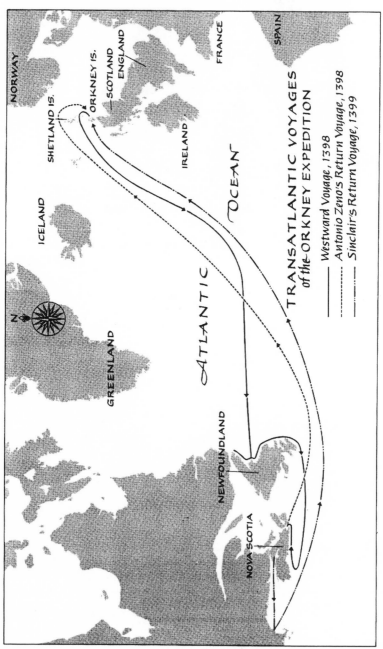

Frederick Pohl's map of Sinclair's transatlantic voyages.

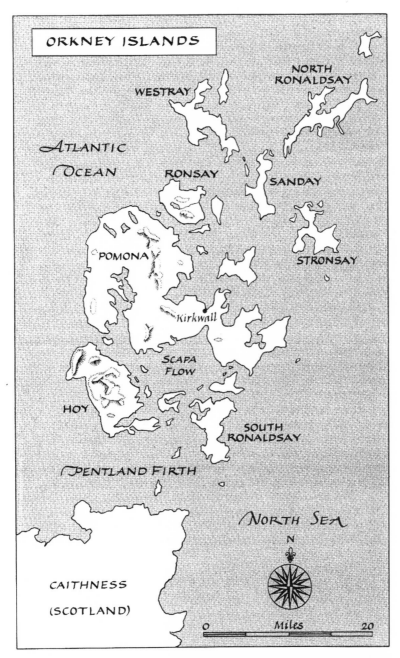

Frederick Pohl's map of the Orkney's and the location of Kirkwall Castle.

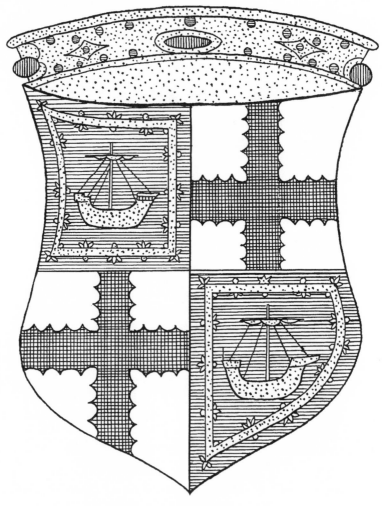

The Sinclair family coat of arms.

William Sinclair and his signature.

7.
Christopher Columbus: Secret Templar Pirate?

You don't change the course of history by turning the faces of portraits to the wall.
—Jawaharlal Nehru

The genius of our ruling class is that it has kept a majority of the people from ever questioning the inequity of a system where most people drudge along, paying heavy taxes for which they get nothing in return.
—Gore Vidal

Was Christopher Columbus actually an agent for the Knights Templar? Had he been a pirate before working for the Spanish crown? Had he assumed the identity of an Italian wool merchant to hide his real identity? Was he a red-haired Jewish sea captain who had been to Iceland? Had some of his Portuguese crew been across the Atlantic before? Is it possible that Columbus knew perfectly well that he would reach a New World by traversing the Atlantic, rather than negotiating a new route to China, which was the purpose he gave the Spanish royalty for his voyage?

For anyone who learned about Christopher Columbus in social studies class, the answer to all of the above questions would be a resounding 'No!' But the real Christopher Columbus may

185

have been an entirely different sort of person from the one we all learned about in school.

Columbus was a fascinating person, and the more one studies this incredible man, the more fascinating he becomes. The man who sailed in the ocean blue in 1492 was a mystic, an adventurer, a utopian idealist and a man of great courage and vision.

The Mysterious Identity of Columbus

Despite the popular grade school depiction, there is a great deal of evidence that Columbus was not actually Italian, but a Spaniard who had assumed the identity of a young Italian wool merchant whom he had once met and who had died at sea. By coincidence, the young Italian wool merchant from Genoa had the same name as himself (at least in translation).

According to Wilford "Andy" Anderson[66] and many other Columbus researchers, we must examine two parallel lives—one of Cristoforo Colombo, born in the latter half of 1451 to Dominick and Susana Colombo in Genoa (Italian "Genova") in the Liguria section of Italy, and the other of Cristobal Colón, born in mid-1460 to Prince Carlos (Charles IV) of Viana, and Margarita Colón in the Jewish ghetto of Genova on the Spanish island Mallorca. The village of Genova is now a district of Palma, the island's principal city.

Contrary to our popular school education, the various details of Cristobal Colón/Columbus' life are not at all well detailed, and very few facts have come down to us through history. Generally speaking, the modern authority on Columbus is Samuel Eliot Morison's *Admiral of the Ocean Sea*[67](1942). Morison was a Ph.D., Pulitzer Prize winner, Book-of-the-Month Club author, an admiral in the U.S. Navy, and a Harvard professor. Morison's is the traditional account of Columbus as the son of an Italian weaver who believed the world was round, convinced the queen of Spain to fund an expedition, and discovered America after thousands of years of isolation—just as we learned in our sixth grade history class.

Impeccable as Morison's credentials seem, Anderson points out that his research was not highly regarded by his peers. "He was ridiculed by his more erudite critics, such as in *National Review* for

April 11, 1975: '*The European Discovery of America*, like so many of Morison's other books, is a flawed work ... marred by a penchant for faulty conclusions, pedagogic prejudices, and bad translations.'"[66] But Morison is not the only one who holds the traditional view of Columbus; hence the profundity of the debate over his life story.

Many of the so-called facts in Columbus' life just don't add up, starting with the navigator's place and date of birth.

Says Samuel Eliot Morison: "There is no mystery about the birth, family and race of Christopher Columbus. He was born in the ancient city of Genoa sometime between August 25 and the end of October, 1451."[67]

However, Simon Wiesenthal, the famed researcher who has sent hundreds of Nazi World War II criminals to their just desserts, gives evidence, as do other authors, that the origin of the Columbus so famous in American history is in Spain. Says Wiesenthal in the book *Sails of Hope*: "He is one of history's most controversial and shadowy figures, with mystery surrounding his birth, his character, his career and his achievement."[68]

According to the Universal Jewish Encyclopedia: "His place and date of birth, generally described as Genoa, in 1446 or 1451, are sharply disputed. ... Most historians have been forced, in great measure ... to reconstruct this obscure period in the life of Columbus on the basis of local Genoese records referring to the Colombo family, who are assumed to be identical with the family of the later Spanish admiral."[70]

But this could be a very bad assumption. One of the inconsistencies in Columbus' life is that he took part in the bombardment of Genoa, his supposed native city, in 1476! Says Morison: "There is no more reason to doubt that Christopher Columbus was a Genoese-born Catholic Christian, steadfast in his faith and proud of his native city, than to doubt that George Washington was a Virginian-born Anglican of English race, proud of his being an American."[67]

Says the Encyclopedia Britannica: "The fact that in the battle (i.e. of August 13,1476) he fought on the Portuguese side, against Genoa, shows him to be no Genoese patriot... One explanation...is that Columbus came from a Spanish-Jewish family settled in

Genoa."[69]

Anderson points out that the family of the Italian Columbus was quite poor, yet Morison asserts: "Domenico Colombo was not a journeyman weaver dependent on wages, but a master clothier (to use the old English term), who owned one or more looms."[67]

Says Wiesenthal: "His father, Domenico Colombo, is supposed to have been a tower sentinel in Genoa and later a weaver in Savona. The family just managed to sustain itself by manual labor."[68]

The Italian Columbus was poorly educated. Says Morison: "One thing is certain, he had little if any schooling."[67] Conversely, says Wiesenthal: "[Columbus] had an excellent command of Latin and Spanish...was well informed in history, geography, geometry, religion, and religious writings...Columbus's whole bearing as a mature man belies the notion that he came from people of small means and had no more than an elementary education behind him."[68]

There are many other points of debate, but, despite the dogmatic assertions by Morison, the contradictions point to the obvious answer that there *must* have been two men. For that we go to two obscure publications of Brother Nectario Maria of the Venezuelan Embassy in Madrid: *Juan Colón the Spaniard* published by the Chedney Press of New York (now defunct) in 1971, and *Cristobal Colón Era Espanol y Judio*, ("Christobal Columbus was Spanish and Jewish") privately printed in Madrid in 1978.

Nectario presents evidence that Colón was the illegitimate son of Prince Carlos (Charles IV) of Viana, Spain, and Margarita Colón, of a prominent Jewish family in the ghetto of Mallorca. The author found a letter from the prince to the governor of Mallorca, dated October 28, 1459, describing his meeting with Margarita, from which can be deduced the birth of Cristobal in the summer of 1460.

This premise provides logical answers to certain "mysteries" concerning Columbus' education and marriage, and reconciles with generally-accepted facts:

1472-73— Colón was in the crew of the pirate Rene d'Anjou on the Mediterranean Sea.[69] It would not be unusual for a boy to run away to sea at 12, especially from a fatherless home. Prince Carlos died in 1461 under suspicious circumstances; some scholars believe he was poisoned on orders from his stepmother.

1473-74— Colón sailed to the Greek island of Chios. The Britannica asserts: "Columbus must be believed when he says he began to navigate at 14."[69] A capable boy of that age with two years of experience would likely be given an occasional turn at the wheel.

1476— He fought with Casenova-Coullon (possibly a relative) against Genoese ships. When his ship caught fire, he swam to the southwest tip of Portugal, near the seamanship academy of Prince Henry, the Navigator,[69] who is believed by some historians to have reached America around 1395.[60]

1477— He sailed to England, Ireland and Iceland.[69]

1478— Married in Portugal to Filipa Moniz Perestrello, daughter of Batholomew Perestrello who sailed for Prince Henry the Navigator and discovered the islands of Madeira and Porto Santo.

1478-83— Colón and his wife lived with her brother, Bartholomew Perestrello II, who had inherited the captaincy of the island of Porto Santo in the Madeiras. From this base, Colón made several voyages to the Gold Coast of Africa.[69] He acquired the papers of Bartholomew I as well as documents from another visitor to the island, Alonso Sanchez de Huelva, who will be discussed later. Colón gained much of his navigational skill during this time, as well as broadening his self-education—the earliest notation in his voluminous library is dated 1481. Morison concedes that "Columbus's exact movements during the eight or nine years that he spent under the Portuguese flag can never be cleared up."[67] Records of Colón are sketchy; records of Colombo end abruptly with his death in 1480.

1484— Colón went to Portugal, where his petition was rejected by the king, probably because there is evidence of a voyage to America in 1472 by the Portuguese mariner Jaoa Vaz Cortereal (known also as Telles) who had a Norse pilot, Pothorst (Jan Skolp).[52] Colón's wife had died during this period, so he left his young son Diego at the monastery at La Rabida, near Palos, Spain, and went to live with Luis de la Cerda, Count of Medina Celi, at Puerto Santa Maria, for the next two years.[69]

1486— Met Ferdinand and Isabella, with the help of the Jewish bishop and professor of theology at Salamanca, Diego de Deza.[70] At about this time, he became involved with Beatriz Enriquez, to whom was born their son Fernando on August 15, 1488.

1489— He was granted the privileges of being lodged and fed at public expense.[69] The next two years are "conjectural" according to Morison. They may have been spent with Beatriz, with his son Diego at La Rabida, or with the Count of Medina Celi.

1491— He appears at Cordoba, where his petition to Ferdinand and Isabella is successful, largely through the efforts of Luis de Santangel, financial minister to the king.

Some of the records of the Italian Colombo are in direct conflict with those of Colón listed above; other records show the probability that the two men met in Portugal or in Porto Santo—or both. Of the "fifteen or twenty notarial records and municipal records" concerning Colombo found by Morison, these are a few:

October 31, 1470— Genoa, a wine purchase is recorded by Domenico Colombo and his son Cristoforo, "over age 19." For Cristoforo, born in 1451, the age matches. Colón, born in 1460, would have been just 10.

March 20, 1472— Colombo witnessed a will in Savona.

August 26,1472— Colombo purchased wool in Savona. Colón was said to be crewing on a pirate ship in this year.

August 7, 1473—Colombo records the sale of a house in Genoa. Colón had set sail for the Greek Isles.

1474— Lease of land in Savona by Colombo, occupation of lessee shown as "wool buyer."[67]

After that, Morison found no trace of the family for several years. But Nectario was able to find records of several wool-buying trips to Portugal by Colombo during the years 1475-78. Colón was also in Portugal in 1478. In 1479, Colombo went to the Madeiras to buy sugar (which was also noted by Morison). Nearest to Portugal in the Madeiras is the island of Porto Santo, where Colón was living at the time. Colombo would certainly have reported to the captain of the island, Bartholomew Perestrello, Colón's brother-in-law. Nectario reports that Manuel Lopez Flores records the death of Colombo at sea the following year, 1480. (There is no evidence that Colombo was pushed overboard by Colón!)

Aside from the obvious difficulty presented by the record of Colombo's death, the other records showing him buying wool and sugar well into his twenties leave us to wonder where and when he could have gained the navigational skills so obviously important to the historical Columbus. Acceptance of the Spanish Jew Colón as the historical Columbus provides logical answers to two major mysteries that have puzzled scholars for centuries:

(1) How had he managed to marry a woman of such prominence? Difficult for a penniless Italian weaver and wool merchant, it would have been no problem for the son of a Spanish prince, legitimate or not;

(2) Where did he obtain funds for his broad education, then available only to the wealthy who employed tutors? Such funds would have been available to the son-in-law of a wealthy family, and some might have come from the royal treasury—there are records of him being supported at public expense from 1489, as previously noted.

The Further Evidence

Once an idea has become entrenched, it is extremely difficult to change because certain groups have a vested interest in sustaining the status quo. In this case, Italians and Catholics enjoyed years in the limelight, when Columbus' world-changing first transatlantic crossing was celebrated every October. Also, countless historians accepted and promulgated the myth. That the facts don't fit the case often matters very little to these groups. But a Spanish-Jewish origin for Columbus is supported by several well-known facts. He invariably wrote in Spanish, and in some letters he referred to Spanish as his "mother tongue." He spoke fluent Spanish, apparently without foreign accent. He apparently spoke no Italian whatsoever. He was able to meet and obtain vital aid from several prominent Jews: Diego de Deza, who arranged for his introduction to their Spanish majesties; Abraham Senior and Isaac Abravanel, who had considerable influence at the court; Gabriel Sanchez, royal treasurer; Juan Cabrero, royal chamberlain; and Luis de Santangel, financier of the voyage as mentioned above.

Probably a quarter to one-third of Columbus' crew members were Jewish, including several of the more prominent, for example

Bernal, the ship's doctor, and Marco, the surgeon. Abraham Zacuto provided astronomical tables which saved the lives of Colón and his crew on the fourth voyage. By the use of the tables, he was able to predict the eclipse of the moon on February 29, 1504, which so awed their Indian captors that they released Colón and his men unharmed. Aboard also on the first voyage, was a Hebrew interpreter, Luis de Torres, who was probably the first man ashore, and who initially reported the use of tobacco by the Indians. Noticeably missing from the ship's roster was a Catholic priest.[68]

Says the *Universal Jewish Encyclopedia*, "Columbus' indebtedness to Jewish scientists, financiers and statesmen forms a remarkable chapter in his career. These personal relationships were all the more significant because of the great role which Jewish wealth, confiscated to the Crown, played in helping finance Columbus' various expeditions of discovery. The wealth was derived from Marranos who were the victims of the Inquisition, and from the Jewish population that was expelled from Spain, en masse, in 1492..."[70]

Colón signed his name with the Spanish form "Cristobal Colón" and in his will insisted that his descendants not vary the signature. He often employed the cryptic form:

S
SAS
XMY
Xpo ferens

This was interpreted by Maurice David in *Who Was Columbus?* as Hebrew for "The Lord, full of compassion, forgiving iniquity, transgression and sin." In letters to his son Diego, in the upper left corner, regularly appear the Hebrew letters *beth he,* for "be'ezrath hashem" (with the help of God) which pious Jews employ to this day.[66]

Colón also marked up the margins of his many books with Spanish notes. He assigned Spanish names to islands he found, such as San Salvador, Punta Lazada and Punta de la Galera: there is no record of his assigning Italian names to any islands. He is known to have been an accomplished mapmaker, primarily a Jewish talent where the activity was centered in Mallorca. And have we mentioned that Mallorca was a major pirate center in the Mediterranean? Perhaps that would explain how the young Colón ended up on a pirate ship at the age of 12.

Peter Martyr, also known as Father de Angliera (or Angheria), wrote the first biography of Colón and is credited with coining the term "New World." Born in 1457 near Lake Maggiore in northwestern Italy, 100 miles north of Genoa, he interviewed Colón extensively after the first voyage. He would have quickly determined that Colón was *not* a native of northern Italy, but would naturally be reluctant to deprive an alleged countryman of his honors by publicizing the fact. He did, however, reveal it in a letter to a close friend, Count Giovanni de Borromeo, who in 1494 left the deposition found in the binding of a book purchased a few years ago from a street vendor in Milan. (A copy is said to be in the library of Barcelona University—the original is held by the family of the count.) The letter reads in part:

> I, Giovanni de Borromeo, being forbidden to tell the truth that I have learned as a secret from Senor Pedro de Angheria, Treasurer of the Catholic King of Spain, must preserve for history the fact that Christobal Colón was a native of Majorca and not of Liguria... He had been advised to pretend, for political and religious reasons, in order to request the help of ships from the King of Spain. Colón, after all, is the equivalent of Colombo, and there has been found living in Genoa one such Cristoforo Colombo Canajosa, son of Domingo and Susana Fontanarossa, who should not be confused with the West Indies navigator.[66]

The Borromeo family was very prominent, boasting (if that is the right word) relations with the de Medicis and Borgias. Carlos Borromeo, likely the grandson of Giovanni, was actually sainted. There is still the possibility that the letter was forged. However, until a battery of experts has submitted it to the necessary scientific tests, the letter fits neatly into the jigsaw puzzle that has baffled historians for centuries.

Columbus and a Transatlantic Refuge for Spanish Jews

Assuming then that Colombo and Colón were certainly different men, how did this colossal case of mistaken identity occur? It is quite likely that Colón himself orchestrated the confusion. After meeting the young Italian wool merchant in Portgual (on the mainland or on the island of Porto Santo), Cristobal Colón the Spanish-Jewish sea captain decided to use the Italian's identity to be able to deal with the Spanish Court. It should be remembered that Spain had only been liberated from the Moors in early 1492. Because of the close collaboration between Jews and Moslems in Spain (and in Morocco and all over North Africa), the Jews were under persecution. In fact, all Jews were officially expelled from Spain, and the very day they were to have left Spain under penalty of law was the day that Columbus set sail for the "New World."

It would have been a very tricky time for Colón. Having utilized the Jewish network to gain financing and favor with the crown, he apparently still felt the need to present himself as a non-Jew. As the *Universal Jewish Encyclopedia* noted, many Jews fell victim to the Inquisition, and more were stripped of their wealth in the mass deportation. It may have been disastrous to his plans had it been known to the Spanish monarchs that Colón began his days in a Jewish ghetto, pirating with a French sea captain. As related by Anderson, "The biography of his son Fernando ... referred to the cover-up in a curious *non sequitur*: 'As God gave him all the personal qualities for so great an undertaking, he wanted to have his country and origin more hid and obscure.'"[66]

That Columbus was a Spanish Jew also gives new interpretation to his reasons for wanting to sail across the Atlantic. Colón set sail on August 3, 1492, the very day that Jews had been banished from Spain. To beat the deadline, he had his crew report on board at 11:00 P.M. on the 2nd, contrary to custom which permitted sailors on long voyages to spend the last hours with their families. Coincidentally, the expulsion date was the anniversary of the second destruction of the Temple of Jerusalem. An estimated 300,000 Jews were banished from Spain in advance of the August 3, 1492, deadline. [68]

Was Columbus looking for a new homeland for the exiled Spanish Jews, or perhaps searching for a lost Jewish city or kingdom on the other side of the Atlantic? In *Sails of Hope*, Wiesenthal states very forthrightly that the main purpose of Columbus' voyage was to resettle the Jews in "Asia" (called "India" at the time), known to have been a haven for the Jews for hundreds of years.[68]

In the twelve volume set, *Life of Christopher Columbus*, it says that Columbus' son wrote of his father: "Their progenitors were of the Royal Blood of Jerusalem." (Vol. 12, page 2). This statement makes one think that Colón was deeply concerned with ancient Jewish history. The Knights Templar took their name from King Solomon's Temple. And, the most famous story of King Solomon is his many voyages to the Land of Ophir to gather the gold to pay for the building of his famous temple.

As we touched upon in Chapter One, it seems likely that the land of Ophir was off to the east of the Holy Land, but there has been much speculation through history as to its location. It has been placed in both North and South America, from New Mexico to the mouth of the Amazon. As a Jew and a mariner, it is certain that Colón would have speculated on this subject himself. He would also know the stories of the Jews seeking refuge from persecution who sailed from Rome in 734 AD "…to Calalus, an unknown land."[75]

Columbus would have known about the legends of the Seven Cities of Gold across the Atlantic as well. The story of the Seven Cities of Gold goes back to 8th-century Spain and the year 711 when the Moorish forces under their general Tariq descended on the Visigoth kingdom which had once dominated the western area of the late Roman Empire, but which, since its defeat by Clovis the Frank in 507, had been reduced to a feeble remnant of its former glory. King Rodrigo rode to battle at the head of his troops, but he was killed and his forces defeated in the battle of Guadalete. The next decade-and-a-half were a horror. The Moorish forces overran all of

present-day Spain and Portugal up to the Pyrenees. Christian refugees escaped in every available direction, often by ship. There is every reason to believe that some Christian refugees left by sea to the Canary Islands, the Madeiras and even to the Caribbean and Florida.

There arose a legend, which persisted throughout the Middle Ages, of seven Portuguese bishops who managed to escape by ship, with a considerable number of the people of their dioceses, and to reach an island somewhere in the Atlantic where they established seven cities. Part of the legend was that the people of the Seven Cities would one day return in force to help their Spanish compatriots defeat the Moors.[72,76]

The legend of the Isle of Seven Cities was kept alive in Spain, and apparently became known elsewhere. In the 12th century the Arab geographer Idrisi spoke of an Atlantic island of Sahelia which once contained seven cities until the inhabitants killed each other off in civil wars.[76]

By the late 14th century, theoretical locations of the Seven Cities were beginning to show up via Spanish and Italian maps on various "imaginary" islands in the North Atlantic. Sometimes the Seven Cities were shown in Brazil but more usually on Antillia, a large island depicted on the maps as opposite Spain and Portugal. According to the map historian Raymond Ramsay, a French map of 1546 appears to be the first to have placed a specific island of *Sete Cidades* in the Atlantic, where it remained on many maps for centuries.[76]

The legends of the cities probably spurred a certain amount of Atlantic exploration, and there were rumors during the 1430s and 1440s of a couple of Portuguese expeditions which were blown off course in the Atlantic to end up at the Isle of the Seven Cities, where the people still spoke Portuguese and asked if the Moors were still in control of their ancestral land. In each case, the rumor had it that some sand from the beaches of the isle was brought home and proved to be rich in gold.[72, 76]

There is a more authentic record of a Fleming whose name is given as "Ferdinand Dulmo," who in 1486 requested the permission of King João II of Portugal to take possession of the Seven Cites, but there is no record that anything was done about it.[72, 76]

Was there a specific island in Caribbean that was the legendary island of Christian refugees? If so, which island was it? The word Antilles is a Latin derivative which means merely "Opposite Island." Ramsay believes that the first time the word was used was the Pizigani brother's map of 1367, where an island labeled "Atilae" appeared in the approximate position of the Azores, which at the time had not been officially discovered. The word appears again as "Attiaela" on an anonymous Catalan map of about 1425 and then on Battista Beccario's map of 1435 where it is spelled Antilia or Antillia from then on.[76]

European geographers apparently took it for granted that a large island was out there on the far side of the Atlantic and the Martin Behaim globe of 1492 showed Antillia with a notation that in 1414 a Spanish ship "got closest to it without danger." This would seem to imply that the island was known, and regarded as a sailing hazard even earlier than that. There are also references to a Portuguese voyage of the 1440s as having reached Antillia, but this is most likely the same one which was supposed to have visited the Seven Cities.[72, 76]

Columbus himself, while in Lisbon in about the year 1480, wrote a letter to King Alfonso V in which he mentioned "the island of Antillia, which is known to you."[76] Columbus also had a letter from the well-known Italian geographer and physician Paolo Toscanelli, dated 1474, recommending Antillia as a good stopping-point to break his voyage to the East Indies and take on supplies.

When Columbus reached the islands of Cuba and Hispaniola, it would seem to fit the description of the legendary landfall of Antilla. Was he in fact, sailing directly for them?

Columbus' friend and first biographer, Peter Martyr, said in 1511 that Columbus believed himself to have found the land of Ophir "but, the descriptions of the cosmographers well considered, it seemeth that both these and the other islands adjoining are the island of Antilla."[72, 76, 59] After Columbus discovered abandoned gold mines on the island of Haiti (Hispanola), he wrote that he had taken into possession for their Spanish Majesties "Mount Soporo (Mt. Ophir) which it took King Solomon's ships three years to reach." Here we see that Columbus had no problem believing that the gold mines on Haiti were worked by ancient Jewish and Phoenician seafarers circa 1000 BC.

In his excellent book on forgotten places and old maps, *No Longer On the Map*,[76] the historian Raymond H. Ramsay says that Columbus' relative Bartolomé confirms the identification of Hispaniola and that Columbus believed himself to have found the land of Ophir, from which King Solomon's ships had fetched gold for the holy Temple in Jerusalem. Ramsay quips that this belief on Columbus' part is of special interest to all students of American history—Columbus' own conjecture *is the first post-Columbian theory of a pre-Columbian discovery of America.*

Given the above discussion, it starts to become evident that Columbus had a certain amount of knowledge of land across the sea before he set out to cross it. The question (as it so often seems to be these days) is, how much did he know, and when did he know it?

The Amazing Chart Room

According to a number of authors, including Charles Berlitz,[77] Columbus was a veritable treasurehouse of navigational knowledge. Says Berlitz, "A dedicated student of earlier Atlantic crossings, he once described the story of two dead men, dark in color and perhaps Chinese, found floating in a long, narrow boat washed up on the western shore of Ireland, near Galway."[77]

Says Anderson, "His 'bravely sailing out into the uncharted sea' is preposterous. On his well-known trip to Iceland in 1477 he would surely have heard of Norse voyages over the centuries ... He had access to a multitude of maps, as extensively documented by Enterline in *Viking America*. Norse author Prytz (*Lykkelige Vinland*) shows that his "Admiral's Map" was almost certainly based on the famed 1424 map drawn in Venice, now in the library of the University of Minnesota. He was born and grew up in Majorca, a center for smugglers and pirates, and for mapmakers, including his own brother Bartholome."[66]

Anderson corresponded with Kare Prytz, a Norwegian newsman whose research in Spain, Portugal and Italy in the early 1970s turned up a large number of original documents showing European knowledge of the early Norse expeditions. In a letter dated

August 17, 1974, Prytz had this to say:

> My conclusion about America's rediscovery is this: Vinland voyages became fimilar in Europe in 1070 via Adam of Bremen, and in the next hundred years the whole of Europe was familiar with these trips. A series of maps were drawn with the Bahamas, Cuba, Haiti, Puerto Rico, etc. ... but the mapmakers didn't know the correct latitudes, so they placed these islands west of Ireland. The name "Vinland" was written "Binini" or "Isla Bini" and was well known long before Columbus. In the meantime, Columbus obtained old maps (especially a map prepared by Andrea Bianco in 1436) and went to Iceland in 1477 for more detailed indications. Here he received America's correct latitude and longitude, and a number of place-names of the stretch from the Florida Keys to Boston.
>
> Columbus know exactly where he should look for the islands and the land he sought, and held a direct course. Orders he gave to the captains were that they were to sail 2,800 miles toward the west, where they would find islands and a continent. After 2,800 miles they were not to sail at night, because they were approaching Vinland. [66]

Remember here that the famous Piri Re'is map states in the legend that it is redrawn from other charts, including maps once used by Columbus. This map is now at the Topkapi Museum in Istanbul and it testifies that the Atlantic was being navigated long before the European Dark Ages. Dating from 1513, the Piri Re'is map shows the entirety of the east coast of North and South America, plus a good portion of Antarctica, just a few years after Columbus made his first voyage to the New World.

Says the Turkish Admiral Piri Re'is in a handbook for sailors called the *Kitab i Bahriye*, Columbus was inspired by a book containing information about these lands, which were claimed to be rich in all sorts of minerals and gems. "This man Columbus tried with the book in his hand to convince the Portuguese and Genoese that an expedition would be very worthwhile. His ideas were rejected and so he turned to the Spanish *bey*. Here too his first request was not granted, but was later on accepted after the matter had been pressed."[73]

And what was the book that Columbus possessed? Admiral Piri Re'is tells us that the book was *Imago Mundi* by the French cardinal Pierre d'Ailly (also known as Aliaco, or Petrus Alliacus), a philosopher, astrologist and cosmographer at the University of Paris. It is known that he had put forward the suggestion that there was land beyond the Atlantic, which he based on his belief that the world was round.[73]

Armed with the *Imago Mundi* and several maps of the Atlantic, Columbus was able to convince the Spanish leaders of the usefulness of an expedition to "India." But amazingly, these tools were not necessarily his most valuable resources. It is fairly well established that he travelled with someone who had already made the journey across the Atlantic.

Nectario found a letter to the Spanish rulers indicating that one Alonso Sanchez de Huelva left his hometown port of Huelva on May 15, 1481, in the ship *Atlante* with a crew of sixteen. It stated that the vessel reached the island of Santo Domingo (called "Quisqueya" by its inhabitants) and on its return stopped at Porto Santo in the Madeiras, where its captain lived for a time. After his sudden death, the ship's papers were given to Cristobol Colón, who was helping run the business of his brother-in-law, captain of the island. It is difficult to believe that such a sensational document is authentic and has been kept a secret for 500 years. However, after some scrutiny, de Huelvo's story has been borne out and it is generally accepted that his ship was blown off course across the ocean more than 10 years before Colón sailed.

The most amazing thing about the letter, however, is it identified de Huelva's first mate as Martin Alonzo Pinzon. A wealthy shipowner of Palos, Pinzon was responsible for Colón's acquisition of the Niña and the Pinta, and for hiring the crew for the expedition. Garcia Fernandez, steward of the Pinta, stated that "Martin Alonzo...knows that without his giving the two ships to the Admiral he would not have been where he was, nor would he have found people, because nobody knew that said Admiral, and that by reason of the said Martin Alonzo and through his said ships the said Admiral made the said voyage."[66] Some historians suggest that Pinzon was the actual leader of the expedition to the New World and that Colón was merely a "front" because of his favor with the Spanish monarchs. As later captain of the Pinta, Pinzon was second in command to Colón.

The steward's letter reconciles with accepted facts. Pinzon's descendants tried for half a century to obtain some of the honors and wealth which had gone to Colón. In the series of lawsuits recorded it was disclosed that Colón had access to the Vatican Archives, which contained records of Viking voyages over a period of several centuries, probably including the original of the famed "Vinland Map" of which Yale University has a 20th century copy. It also indicated that Pinzon had obtained from Rome information about lands to the west of the Atlantic.[66]

In his capacity as advisor on the voyage, Pinzon was able to get Colón to alter his course on October 6th.[67] What better reason would the Admiral have had for accepting the urging of a subordinate (especially one with whom he had a competitive rivalry) than the fact that Pinzon had been there before?

The Templar Connection

That Columbus was more than he initially appears would fit into the strange history of maritime activity in Europe and the Mediterrean following the disappearance of the Templar fleet. His familiarity with all the normal spheres of activity at the time, plus a few more far-flung, would have brought him into contact with all sorts of people. His apparent Jewish ancestry, and his confident knowledge of the transatlantic world would point to his being a consort of the Knights Templar and their allies in Portugal and Spain.

Close your eyes for just a moment and picture the drawings of the Nina, Pinta and Santa Maria you saw in your history books. Remember the square sails? Remember that they were white with red crosses emblazoned across the middle? These were the ships of the Knights of Christ of Portugal, the Templars by another name.

Michael Bradley says that Columbus was said to have blue eyes, a pale complexion and light colored hair. Appearance, of course, does not necessarily mean anything, yet this complexion is not what we normally expect of Mediterranean Jews, or Sephardic

Jews, who came into Iberia from Palestine during the Diaspora. Some people said that Columbus's hair, blond in his youth, had a distinct reddish tinge in later life.[9]

If appearance means anything at all, the description of Columbus suggests a Celtic or Nordic genetic heritage, not a south European one. He could have come from the Celts of the Pyrenees Provencal culture; or from Brittany and Anjou where Viking raids of the 8th-11th centuries had left a Nordic genetic legacy. Or he could have come from Sicily, which the Norse king Roger (1130-1154) had conquered with his many Nordic vassals and noblemen. Or was he from Mallorca, which also acquired a Nordic-looking aristocracy when Roger's ancestors decided to take that island?

In private correspondence to Bradley, English historian Michael Baigent, co-author of *Holy Blood, Holy Grail*,[1] said he tends to think that Columbus may have been born on Mallorca where, just behind the Templar castle in the capital town of Palma, there was a district in which resided a number of cartographers recruited from all over the known world. Was there still a Templar influence in the area, too?

After a while, criss-crossing coincidences become probabilities. And then there are the just plain facts that finally come into focus. So, revisiting the questions from the beginning of this chapter, it begins to look like the answers are 'Yes!'

Christobal Colon-Colombus

The cryptic (Templar?) cipher that Colombus always used with his signature.

Columbus's house in Porto Santo, near Madeira. It was here that his elder son Diego was born, probably around 1480.

The ruins of the house in which Columbus is said to have lived for a time in Funchal, Madeira.

With the aid of an egg Columbus demonstrates to Spanish skeptics that a man can do anything if he knows how. Engraving by Theodore de Bry.

Theodore de Bry's allegorical engraving, *The Vision of Columbus*.

A 1494 woodcut of the arrival of Columbus in the New World.

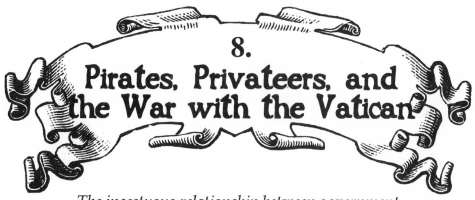

8.
Pirates, Privateers, and the War with the Vatican

The incestuous relationship between government
and big business thrives in the dark.
—Jack Anderson

The pyrates... are unanimously reputed to be English.
—East India Company Official

Today, piracy has taken on different connotations from the early days of naval war between the remnants of the Knights Templar and the Vatican. As they say in this modern era of national struggles: You are a freedom fighter or a terrorist, depending on whose side you are on and who wins the war.

In those early days, the war between the Vatican and the lost Templar fleet meant that the Templar ships flew their flags with pride: the Jolly Roger and other Masonic-Templar flags. Some Templar ships, like those in Portugal, still flew the Templar flag. These, indeed, were the ships that Christopher Columbus sailed in for the Spanish crown.

§§§

The Papal Line of Demarcation of 1493

As soon as news of Columbus' successful voyage reached Europe, the Spanish monarchs moved to secure their new trade route. Since the days of Prince Henry the Navigator, Portugal had been actively exploring and developing sea trade with new lands. Throughout the 15th century, Portuguese expeditions reached farther down the coast of Africa and into the Atlantic. The interests of Spain and Portugal now collided head-on.

Since both countries were Catholic, the Pope was petitioned to decide the territorial conflict, as had been done in the past. In this case, the Pope was Alexander VI. Born Rodrigo de Borja (Borgia) to a prominent family of Valencia, Spain, he was the nephew of Pope Callistus III. Rodrigo had several illegitimate children by his mistress Vanozza Cattani. Upon his installation as Pope, his son Cesare Borgia took over the direction of his affairs. The political maneuvering, corruption and greed of this administration was unequaled. Moral laxity and unconcern for the spiritual guidance of the church were the norm. Cesare is said to be the model for Machiavelli's Prince. Cesare's sister, Lucrezia Borgia, is well known in her own right.[95]

When called upon to settle the dispute between Spain and Portugal, Alexander VI issued a papal bull which divided the world up between the two countries. The Pope maintained that this would advance the cause of Christ. Said the bull, in part:

> Among other works well pleasing to the Divine Majesty and cherished of our heart, this assuredly ranks highest, that in our times especially the Catholic faith and the Christian religion be exalted and be everywhere increased and spread, that the health of souls be cared for and that barbarous nations be overthrown and brought to the faith itself...
>
> We have indeed learned that you, who for a long time had intended to seek out and discover certain islands and mainlands remote and unknown and not hitherto discovered by others, to the end that you might bring to the worship of our Redeemer and the profession of the Catholic faith

their residents and inhabitants...[96]

Since there is evidence that the recent Spanish voyage did not even include a Catholic priest, the motivation of the countries in the dispute could be better interpreted as to discover new lands and usurp their riches. Perhaps this was a tacit understanding, since the Pope would certainly see his share. Unsurprisingly, the Pope's decision slightly favored Spain. Basically, a line was drawn which ran from pole to pole 100 leagues west of the Azores. Anything to the east of it belonged to Portugal; anything to the west belonged to Spain. (See the map in the picture section). This was appealed by Portugal, and the line was moved further west— which had the "unintended" effect of landing the easternmost part of Brazil in Portugal's territory, sanctioning the colonization of that land by the Portuguese.

Obviously, the papal bull was utterly politically incorrect vis a vis the rights of indigenous peoples of lands around the world to their own sovereignty and religious practice. But it also consummately ignored the other European sea powers. And for their part, England, France and the Netherlands ignored the papal bull. If anything, it served as a wake-up call to these countries that Spain and Portugal had a leg up on them in the race to leave the Dark Ages behind. The Age of Discovery—and colonization and accumulation of foreign wealth—began in earnest.

§§§

The Origin of Privateering
According to the 1997 *Grolier Encyclopedia*, the definition of privateering is: "Privateering was the practice by which governments employed privately owned vessels to capture enemy commerce during wartime. Motivated by the profits made by such captures, privateers held licenses, called letters of marque and reprisal, and their attacks were limited to vessels of nations at war with the country issuing their commissions. Privateering differed from piracy mainly because of those licenses and limitations. Although privateers existed from at least the 13th century and into the 19th, the heyday of privateering was from the 16th to the

18th century."[95]

Indeed, the origin of privateering was the secret war between the Templars and the Vatican and the Pope's allies. This meant that privateers were mainly English, Scottish, or American ships. Some were Dutch or French, and these privateers were typically Masons. The targets for privateers were almost exclusively Spanish or French ships, or Moslem ships plying the Mediterranean or Indian Ocean.

As the next century or two loped along in Europe and the New World, a dizzying interplay of politics and religion resulted in an ever-changing kaleidoscope of alliances and animosities. Since the French Templar families had left France at the time of the suppression of the Templars, and gone largely to England and Scotland, France had returned to the secure fold of the Vatican.

The Merovingian family of "Holy Blood" had long since ceased to rule, and in the 16th and 17th centuries France was almost continually at war with England. This made French ships, as well as Spanish (and some Portuguese) the target of pirate and privateer attacks. Some pirates, indeed were French, and I like to think that some of these pirates were Templar-Mason remnants who took a certain pleasure in attacking the treasure laden ships and ports in the Vatican's New World.

So, privateers differed from pirates in that they, in theory, had some sort of official commission to go out and attack certain ships of enemy countries. Always difficult to control, privateers often lapsed into piracy. By 1500, international agreements limited their illegalities, requiring all seizures to be confirmed by admiralty courts, and thus distinguishing pirate from privateer. Gradually privateering operations were carefully regulated, and by 1740 privateering was an accepted area of business investment. When national navies were small, privateers could effectively injure an enemy at little cost to the licensing power.

All western European nations, especially weaker naval powers, employed this means of destroying enemy trade. Originating in Europe, it eventually spread throughout the world, reaching great vogue in the Caribbean and along the North American coast.

Later, after Magellan and Drake had circumnavigated the world and the Spanish galleons began plying the Pacific Ocean from

Acapulco to the Philippines, piracy and privateering spread to the Pacific coast of the New World and to Southeast Asia and the Indian Ocean. Madagascar became a major pirate area, a large and ungoverned island that lay perfectly in the sea lanes between India and the Cape route to Europe. Southeast Asia today is still a hotbed of piracy (though they are not in any way a remnant of the Templar fleet) and piracy still exists along the wild Pacific coasts of Ecuador, Colombia and Panama.

§§§

Templar Scotland Revisted

First, let us have a quick recap of Scottish history leading from the Templar arrival starting in 1307 to the creation of the United Kingdom. Scottish kings were often having disputes with the English, which occasionally led to war and ultimately the paying of homage to English kings based in York.

In 1174, for example, William the Lion did homage to Henry II of England for all his dominions, although he later secured cancellation of this obligation. After the deaths of Alexander III (1286) and his heiress, Margaret, Maid of Norway (1290), King Edward I of England claimed suzerainty over Scotland and picked John de Baliol from rival claimants as the Scottish king. When John attempted to assert his independence in 1296, Edward imposed direct English rule.

The Scots, led first by Sir William Wallace and later by Robert the Bruce (crowned king as Robert I in 1306), revolted and finally defeated the English in the Battle of Bannockburn (1314), in which we have said the Knights Templar took part, and Scotland retained its independence from England.

But later, during the long reign of David II (1329-71) the English partially reestablished control, but they were distracted by the Hundred Years' War with France, Scotland's ally since 1295. David was succeeded in Scotland by his nephew Robert II, the first Stuart monarch. He and his successors, Robert III, James I, James II, James III, James IV, and James V faced persistent civil strife and continued interference from—and war with—England.

After the death (1542) of James V, Scotland was ruled by his widow, Mary de Guise, as regent for her young daughter, Mary, Queen of Scots. During this period the Reformation took root in Scotland, gathering strength as a result of the political opposition to the French regent. The Catholic Mary, Queen of Scots, who returned to Scotland from France in 1561, fell victim to the religious and political conflicts. Forced to abdicate in 1567, she fled to England, where she was imprisoned and later executed by Elizabeth I. Nonetheless, her son, James VI of Scotland, succeeded Elizabeth on the English throne as James I in 1603. Although united under a single crown, Scotland and England remained separate states for another century.

The Scottish Presbyterians resisted the efforts of the next monarch, King Charles I, to impose episcopacy in Scotland and ultimately their rebellion brought about the execution of the King in 1649 and a civil war in England. It had now been hundreds of years since the suppression of the Templars and entirely new countries had been created with new alliances. Parts of the Templar fleet which had become the navy of Scotland and its new inheritors, became Masonic English and Dutch privateers who attacked Spanish and French ships that were allied with the Vatican. The Templar tradition infiltrated England and the North American colonies under the guise of Masonry. There were two main forms: Scottish Free Rite Masonry and York Rite Masonry.

§§§

Sir Dragon and the Masonic Lodge of Privateers

What is curious in our history of the lost Templar fleet is that many of the pirates and privateers that were to help in the creation of the United States were also Masons. Though it is impossible to prove that all these men were Masons (it was, after all, a secret society), we know that many of them were.[78]

The Elizabethan wars between Spain and England (and other countries) led to a new outbreak of "pirate" attacks in the 16th century. Galleons carrying treasure from America to Spain were tempting targets for seafaring raiders. These chartered privateers and pirates had names such as the Sea Dogs of England, the Sea Beggars of the Netherlands, and the Sea Wolves of France. They intercepted countless Spanish vessels, including, in 1521, a ship laden with riches from Hernan Cortez's Mexican expedition.

By the late 16th century, tempted by the booty available, some sea rovers—including Sir John Hawkins and Sir Francis Drake—attacked Spanish colonies in America. By the 17th century pirates had established Caribbean bases at Saint Christopher's, Jamaica, Tortuga, and other points. The Spanish dominated the Caribbean and the New World at this time, but it was up to the Templar navy in the guise of Masonic privateers to turn the tide for the Scottish and English and their secret Templar masters.

This is the era of adventure on the high seas that is depicted in pirate movies such as Roman Polanski's *Pirates*, the Geena Davis film, *Cutthroat Island*, and Disney's recent *Pirates of the Caribbean*. It was during this romantic era that Sir Henry Morgan sacked and pillaged along the Spanish Main, and was later the governor of Jamaica (where he tried to control piracy).

In proof that there was some genuine romance to this life of privateering for the Crown of the United Kingdom, even women were attracted to pirate life. Female pirates such as Anne Bonney and Mary Read participated in many coastal raids and had their own "pirate" crews.

The most notable Masonic privateer of all time is Sir Francis Drake. Drake (whose name means "dragon"), was the most famous English seaman of the Elizabethan Age and is best known for his exploits against the Spanish and his circumnavigation of the world. Drake was born in Devonshire about 1541 to a poor, staunchly Protestant farming family. He went to sea as a youth, aided by his relations, the Hawkins family of Plymouth. He took part in a slave-trading expedition to the Cape Verde Islands and the West Indies in 1566.

The next year he sailed under John Hawkins on an expedition that was attacked by the Spanish at San Juan de Ulua (Vera Cruz, Mexico); only the ships commanded by Hawkins and Drake escaped. Drake scouted the Panama region in 1570 and returned there the next year, when he bombarded several coastal cities and captured much gold and silver. Temporary duty in Ireland followed.

The deterioration of relations with Spain led Queen Elizabeth I to back an expedition to sail around the world, a feat that had been accomplished only once before, by a Spanish expedition under Ferdinand Magellan. Drake was placed in charge of five small ships and about 160 men and left Plymouth in December 1577. The voyage was a difficult one. Drake had to put down a mutiny off the Patagonian coast and abandoned two small storeships. It took 16 days to pass through the Strait of Magellan; one ship abandoned the quest and returned to England and another disappeared during a storm. Thus, by the time he reached the Pacific, Drake was left with only one ship, the Golden Hind.

He traveled up the coasts of Chile and Peru raiding Spanish shipping and sailed far to the north before returning south to touch land in the San Francisco area; he claimed the surrounding region for England and named it New Albion. Drake then set off across the Pacific; he traded in the Spice Islands (Moluccas) and signed treaties with local rulers. He returned to England via the Indian Ocean and the Cape of Good Hope, reaching Plymouth on Sept. 20, 1580, laden with treasure and spices. For his accomplishments, he was knighted by Queen Elizabeth I.

Drake served as mayor of Plymouth for some years. In 1585, however, he was given command of a large fleet that sacked Vigo, in Spain, then crossed the Atlantic to capture and plunder Cartagena and Santo Domingo and to destroy St. Augustine, in Florida. In 1587 he led a raid on Cadiz that destroyed the stores and many of the ships of the Spanish fleet. This exploit delayed the attack on England by the Spanish Armada for a year. Drake played a major role in the successful English defense against the armada in 1588 and was acclaimed as England's hero. But his forces were defeated at Lisbon in 1589, and for some years he remained in Plymouth. A final voyage to the West Indies, begun in 1595, was totally unsuccessful; Drake died off the coast of Panama on Jan. 28, 1596.

Drake's cousin, Sir John Hawkins, was another famous Masonic privateer. Hawkins (b. 1532, d. Nov. 12, 1595) was a famous English naval commander of the Elizabethan era. Born into a family of seafarers at Plymouth, he made his earliest voyages to the Canary Islands as a young man. He married Katherine Gonson, whose father was treasurer of the navy, in 1559. After Gonson's death (1577), Hawkins assumed the post and introduced notable improvements in shipbuilding and naval administration.

Between 1562 and 1569, Hawkins led three expeditions in which black slaves were taken from Africa for sale to Spanish colonies in the West Indies. On the third voyage (1567-69), in which his kinsman Francis Drake took part, he was involved in a major battle with a Spanish fleet at San Juan de Ulua, off the coast of Mexico, and lost many of his men and ships. Hawkins commanded a portion of the English fleet that defeated the Spanish Armada in 1588. He died at sea on an unsuccessful expedition to the West Indies; Drake died during the same expedition.

In the next century the most famous Masonic privateer was the Welsh adventurer, Sir Henry Morgan. Morgan (b. 1635, d. Aug. 25, 1688) was one of the buccaneers who, with the unofficial support of the English government, preyed on Spanish shipping and colonies in the Caribbean. In 1668 he captured Puerto Principe (now Camaguey, Cuba) and sacked Portobelo (now in Panama). He raided Maracaibo (now in Venezuela) in 1669.

His spectacular capture of the city of Panama in 1671 was marked by great brutality and debauchery. Afterward much of the booty was lost and Morgan's crew claimed that he had cheated them. Captured and sent back to England in 1672 to answer piracy charges, Morgan was treated like a hero, knighted, and appointed lieutenant-governor of Jamaica, where he lived quietly thereafter.

Such successes against the Spanish colonies in America were praised from tavern to tavern along the coast of New England and the romance and lure of piracy and privateering pervaded the English, Scottish, Dutch and other colonists. There was nothing shameful in the acts of these privateers, in fact they were heroes. But the era of royally commissioned pirates was coming to an end. England was now a powerful nation with a large navy, thanks to the former Templar fleet and the successful union of Scotland and England. Support of privateering begins to look a lot less attractive when you are the one with the most to lose by it.

§§§

Jolly Rogers, Black Flags and the Red Flags

As mentioned in chapter three, pirates often flew a flag known as the "Jolly Roger," named after the Templar king of Sicily, Roger II. The flying of flags, especially on ships, in the Middle Ages was very important. Communication from ship to ship and ship to shore was by flag. With nations at war, and often switching sides and alliances as well, the flying of one flag or another could mean the difference between a passing ship attacking you or not.

The ships of various nations always flew their national flags, often with accompanying banners. Pirate ships were different. They often carried several flags with them of various nations, and flew whichever flag was appropriate for getting close to another ship.

The pirates always had the advantage when approaching a victim. They could follow a ship for hours or days at a safe distance while they worked out her potential strength in terms of guns and crew. If she proved to be a powerful Indiaman or a man-of-war, the pirates could veer away and seek a weaker victim. If the vessel appeared vulnerable, the pirates had a choice: they could take her by surprise, or they could make a frontal attack.

The simplest method of catching a victim off guard was to use false flags, a *ruse deguerre*, which naval ships frequently adopted in time of war. Before the advent of radio or Morse code signaling, the only way that a sailing ship out at sea could identify the nationality of another vessel was by her flags. By 1700 the design of national flags was well established, and an experienced seaman was able to identify the ships of all the seafaring nations by the colors flying at their mastheads or ensign staffs.

A black flag with a white skull and crossbones emblazoned on it has been the symbol for pirates throughout the Western world. However, the Jolly Roger was only one of many symbols associated with piracy. In the great age of piracy in the early 18th century a variety of images appeared on pirate flags, including bleeding hearts, blazing balls, hourglasses, spears, cutlasses, and whole skeletons. Red or "bloody" flags are mentioned as often as black flags until the middle of the 18th century. The Jolly Roger, at least in the early days, meant an association with the lost Templar Fleet.[24]

A French flag book of 1721 includes hand-colored engravings of pirate flags, including black flags with various insignia, and a plain red flag alongside a red pennant. Under the red flags is written "Pavilion nomme Sansquartier" ("Flag called No Quarter"). The idea that a red flag could mean no mercy is confirmed by Captain Richard Hawkins, who was captured by pirates in 1724. He later described how "they all came on deck and hoisted Jolly Roger (for so they call their black ensign, in the middle of which is a large white skeleton with a dart in one hand, striking a bleeding heart, and in the other an hourglass). When they fight under Jolly Roger, they give quarter, which they do not when they fight under the red or bloody flag."[24]

It is interesting to note that the flag that Hawkins describes is a flag that was flown, among others, by the North Carolina pirate Edward Teach, commonly called Blackbeard.

Pirate authors often mention that those who flew the Jolly Roger were not necessarily cold-blooded killers. According to David Cordingly in his book *Under the Black Flag*,[24] pirates weren't such bad guys really, and were often very gentlemanlike in their behavior.

Cordingly says that in the great majority of cases merchant ships surrendered without a fight when attacked by pirates. If a merchant ship surrendered without a fight, the pirates usually refrained from inflicting violence on the crew. Typically, the cargo of the ship was transferred from one ship to another, the crew of the other ship disarmed, and everyone was set free. Food and water were left on the ship, and, as we shall see, pirates treated themselves very democratically, and typically their captives as well.

In the boarding of a ship there were often some casualties, though pirate ships were usually so heavily armed with cannon and desperate freebooters that most ships at first tried to out-run the pirate ship if they could, and then surrendered when captured.

When a ship flying a skull and crossbones or some variant came within sight of a Spanish treasure ship in the Caribbean, there would be curses and screams. Women would faint, men would instinctively reach for their swords, and a race would be on to see who had the faster ship: the pirates with their Masonic flags, or the treasure-ladden ships of Spain.

Pirate ships, because of the natural tendencies of victim ships to attempt to out-sail them, were usually very fast ships. When pirates occasionally captured faster ships than the ones they were currenly using (such as in a port), they would usually keep that vessel for themselves.

According to various pirate books, the most well known pirate who flew the Jolly Roger was called "Captain England." A skull and crossbones with an hourglass beneath them was the pirate flag of the French filibusters, which translated into English means "freebooter," a popular term for a pirate. One French pirate ship, the *Sanspitié*, flew a flag which featured a skull and crossbones plus a naked man holding a sword in one hand and an hourglass in the other. Captain Bartholomew Roberts, an English pirate, had a flag in which a sea captain and a spear-holding skeleton hold an hourglass between them.

All pirate flags were loaded with symbolism, and the meaning of the hourglass is an especially interesting one. One might conjecture that it meant that time was running out for the Vatican and their allies.

Other pirates flags of the time consisted of a skull and crossed scabards. This striking design, used as a chapter header for this book, and in Disney's *Pirates of the Caribbean* film, was originally used by "Calico Jack" Rackham. In fact, Johnny Depp's character in Pirates of the Caribbean was probably modeled after Rackham. He was known as "Calico Jack" because of the shirt and breeches of calico that he habitually wore. Textiles were a valuable and popular cargo on ships, and Rackham was a particularly snappy dresser. He was an educated and handsome man, fearless and resourceful. The two most famous women pirates of the time, Mary

Read and Anne Bonny, both served with Rackham, and Bonny probably had his child.

Rackham's main base of operations was in the Bay Islands of Honduras (portions of which were to become British Honduras, later renamed Belize) but as he became more successful, he established a family headquarters at an obscure port in a small island south of Cuba, where Anne Bonny resided during periods of domestic indisposition, returning to the ship when the "discomfort" was past. He now paid habitual visits to the "family" and stay until money and stores were gone.

In the most famous story of the flamboyant pirate, while he was in his secret port clearing his ship and preparing to depart, a Spanish *guarda costa* came in with an English sloop caught illegally trading on the coast. The Spaniards made bold to attack Rackham, who lay close to the shore of the island and well out of range. Therefore the *guarda costa* sailed into the channel during the evening to blockade the small bay. The Spanish felt with certainty that Rackham was trapped and they could destroy the pirate readily in the dawn.

Rackham soon perceived the desperate situation he was in and acted like a man of courage and resource. The captured British sloop lay in such a position that she could be used for escape. Rackham, knowing that his ship would be blown to smithereens by cannon from multiple ships in the morning, decided on a highly inventive maneuver. Rackham manned his boats in the dark and with muffled oars slipped from his brigantine to the sloop. Quickly mastering the prize crew and making her his own, he cut the cable and put to sea.

In the morning the Spaniards opened with all their guns upon Rackham's brigantine, which to their amazement remained silent under the fire. Their chagrin may be imagined when it was discovered that the birds had flown with a valuable prize, leaving nothing but an empty and well-worn hull behind.

Rackham continued to attack cargo ships and made his new base of operations in Jamaica. In 1720, when the governor of Jamaica learned of Rackham's presence, he sent an armed sloop to capture him. Rackham tried to escape in his ship, but the British warship overtook him, and with barely a fight, Rackham surrendered. Perhaps he thought that the governor of Jamaica would have mercy on this fellow Englishman, but most of the crew were sentenced to be hung. Mary Read, Anne Bonney and some of the crew who claimed they had been pressed into service as pirates, were acquitted. Rackham was hung at Port Royal at the entrance to the port where all seamen might see his body for months to come.

Anne Bonney is reported to have said of Rackham, "If he had fought like a man, he need not have died like a dog."[57, 58]

§§§

Captain Kidd and the End of the Templar Pirates

Of all of the Templar-Masonic pirates, perhaps none has garnered more interest than Captain Kidd. William Kidd was born in Scotland in 1645 and died May 23, 1701. He was a Scottish-British pirate whose life has been much romanticized by literature.

In 1689 William Kidd served on the pirate ship *Blessed William*. The ship surrendered at Nevis, an island in the Caribbean. Kidd was given license to attack the French by the governor of Nevis. In December 1689, he participated in the plunder of Marie Galante as well as attacking several French ships. In February 1690, Kidd's crew fearing for their safety in war decided to steal the ship *Blessed William* while Kidd was ashore. Kidd was given the *Antigua* and gave chase, chasing the *Blessed William* to New York. New York at the time was embroiled in civil war. Kidd allied himself with the winning side and married a wealthy widow in May 1691. He remained at New York for a few years acquainting himself with political leaders. One of those he made acquaintance with was Robert Livingston, who was an ambitious entrepreneur.

In 1695, he and Livingston sailed to London where they met with Richard Coote, Earl of Bellomont and recently-appointed Governor of New York and Massachusetts. Kidd had hopes of securing a privateering license. The three concocted a scheme to capture pirates and keep the booty for themselves instead of returning it to its owners. The three signed a contract in October with Coote staking £6,000 for outfitting Kidd in his expedition.

(Coote made several other arrangements secretly that involved the Secretary of State, the heads of the Admiralty and Judiciary Courts, and the King himself.) Kidd was granted three commissions by the king. The first allowed for the capture of French ships, the second allowed for the capture of pirates everywhere, and the third and most important to their cause, allowed the suspension of all captured booty having to go through the courts. This allowed Kidd to keep the booty until time for surrender of booty to Governor Coote in Boston.

The three bought the *Adventure Galley,* a 300-ton, 34-gun ship. Kidd left England in May, 1696 sailing for New York where he recruited his crew. He promised his crew 60% of all booty even though he had already promised 60% to Governor Coote.

In September, he sailed to the Indian Ocean by way of the African coast. He stopped along the way at Madagascar, careening his ship at Johanna Island (Anjouan). In April 1697, Kidd left Johanna Island and sailed for the Red Sea, planning to plunder pilgrim ships returning from Mecca to India.

Although Kidd may have originally intended to attack other pirate ships that he encountered, as his second commission stated, it was an unlikely scenario. His crew was unlikely to attack their fellow pirates, many of whom were their friends. Furthermore, attacking a ship full of hardened sea dogs was quite different from attacking a merchant ship. Indeed, Kidd would meet the pirates John Hoar and Dirk Chivers at Saint Mary's island in Madagascar, but there would be no attacking them.

§§§

A Contract Between Pirates

One of the documents that has come down to us through pirate history is the contract signed between Capt. Kidd and his pirate crew. This crew, as history likes to tell us in the many popular books written over the last few centuries, was a bunch of scurvy dogs—seasoned pirates who were sent foolishly after other pirates, even more unsavory and ruthless. As their captain was a tall and commanding Scotsman who had settled in New York, he knew pirates, but was now a pirate hunter. The tide had turned on the buccaneers and the war with the Vatican. Now the pirates were outside of all law, including the law of the Knights Templar and the Free Masons that succeeded them.

After the ship had been away from New York for several days, Captain Kidd made a contract with his motley crew. He again reiterated that he was commissioned by the King of England to attack and capture French ships as well as "other pirates."

Kidd's crew included a forty-two-year-old Jewish jeweler named Benjamin Franks, a Ceylonese cook, a native American Indian, and 147 other men who were freebooters or merchants hoping for a rich and easy prize. Captain Kidd's ship was well armed for any ship on the seas at that time, and each of the men could come home quite wealthy.

On September 10, 1697, each of the 150 men came into the captain's cabin to sign the contract—most merely applied an X or perhaps their initials. This fascinating pirate document has survived and it shows the democratic philosophy that ruled such a ship at war:

"ARTICLES of Agreement ... between Capt. William Kidd Command.er of the good ship die Adventure Galley on the one part and John Walker Quarter M.er to die said ships company on the other part, as followeth, vide . . ."

This contract is a kind of negotiated treaty between 150 men on one side and the captain on the other. On a Royal Navy ship, the captain's authority is reinforced by a troop of armed marines; on this privateer, Kidd's power comes mainly from this piece of paper (and the force of his personality and a mere handful of loyal officers).

The articles also give us a general code of conduct for these warriors:

* Incentive Bonus: "The man who shall first see a Saile. If she be a Prize shall receive one hundred pieces of eight."

* Workman's Compensation: "That if any man shall lose an

Eye, Legg or Arme or the use thereof [he]... shall receive ... six hundred pieces of eight, or six able Slaves."

* Discipline: "That whosoever shall disobey Command shall lose his share or receive such Corporall punishment as the Cap.t and Major part of the Company shall deem fit.' (This important clause in the contract says that Captain Kidd could not punish his men without the consent of the majority, a truly democratic ship.)

* Cowardice: "That man is proved a Coward in time of Engagem.t shall lose his share."

* Sobriety: "That man that shall be drunk in time of Engagement before the prisoners then taken be secured, shall lose his share." (A drunken post-victory celebration was typical after a ship was taken.)

* Loyalty: "That man that shall breed a Mutiny Riot on Board the ship or Prize taken shall lose his shares and receive such Corporall punishment as the Capt. and Major part of the Company shall deem fitt." (As before, Kidd rules by democratic vote of the entire ship.)

* Honesty: "That if any man shall defraude the Capt. or Company of any Treasure, as Money, Goods, Ware, Merchandizes or any other thing whatsoever to the value of one piece of eight... shall lose his Share and be put on shore upon the first inhabitted Island or other place that the said ship shall touch at."

* Sharing: "That what money or Treasure shall be taken by the said ship and Company shall be put on board of the Man of War and there be shared immediately, and all Wares and Merchandizes when legally condemned to be legally divided amongst the ships Company according to Articles." (Kidd reserved 40 shares for himself and the owners, with the rest to go to the crew.)

Captain Kidd's signature was a large slashing W and an oversized swirling K. He confidently agreed to what was probably a

standard "pirate" contract. He was their leader, but he agreed to give away his right to order up lashes for his crew unless the majority agreed to it.

No other types of ships had such motley, but democratic crews in those days. British, French, Dutch, Spanish and other merchant and navy ships had hired sailors who worked for a fixed pay. No matter how rich a cargo might be, even the spoils of some sea battle, the sailor's pay would not change. For privateers, freebooters, buccaneers, and pirates, the pay they would receive was on a sliding scale. Depending on their luck, their hard work, their daring—they might be sharing a rich prize full of gold, silk, rum, chests of coins and more.

The contract articles on Kidd's ship show the inherent democracy and equality that was common among privateers and pirates, reflecting the Templar ideals of the early pirates. As a pirate or privateer on such a ship you could be a Christian European, or a Hindu from Malabar, or a Buddhist from Ceylon, or a Moslem from North Africa. You might be an American Indian, and you could be black, brown or white skinned. All were equal partners on the ship (except for the Captain). A pirate's life was much like the burgeoning republic that would become the United States: all were equal on the ship, and no one could be punished without the majority of the crew agreeing to it. As I have already said, this is unlike any other ships that were sailing the Seven Seas at that time.

But Captain Kidd had a problem. He was to attack pirates, but his own crew were pirates themselves, men who owed no allegiance to the English king or any other authority. Governor Fletcher of New York later observed, "When [Captain Kidd] was here, many flocked to him, …men of desperate fortunes and necessities, in expectation of getting vast treasure. He sailed from hence with 150 men ... great part of them from this province. It is generally believed here that they will get money *per fas aut nefas* [one way or another], and that if he misses the design named in his commission he will not be able to govern such a herd of men."[17]

As Richard Zack's says in his recent biography of Kidd, *The*

Pirate Hunter, "So here it is: Captain Kidd's mission is to go chase pirates—men who would rather die than surrender. He is to travel in a lone ship manned with a desperate crew, some of whom are former pirates. His ship's articles do not allow him to punish his crew, except by vote of the entire crew. As a private man of war, he will be deeply distrusted by the Royal Navy; as a commercial rival, he will be despised by the English East India Company. He is a Scot lording it over an English and Dutch crew. Once he rounds the Cape of Good Hope, he will find no welcome ports of call, except pirate ports. On the immense Indian Ocean of twenty-eight million square miles, he must find some of the five currently active European pirate ships, many of them carrying relatives and friends of his crew. And he has a one-year time limit and some of the most powerful men in the world waiting for him to return. It would be a fool's errand—except for the treasure."[17]

§§§

Kidd Captures a Rich Moslem Treasure Ship

On August 15th, Kidd encountered an Indian squadron escorted by an East India Company ship. The British ship opened fire on Kidd. Kidd retreated to northwestern India. On the 19th, Kidd seized a small ship near Janjira. He allegedly tortured the Indian sailors, and impressed the British captain into acting as pilot for several months. Kidd then continued south, fighting off two Portuguese warships and stopping in September at Laccadive Island for repairs. At Laccadive Island, Kidd's crew forced the inhabitants to work, used their boats for firewood, and raped their women. In November, Kidd encountered another East India Company ship. His crew wanted to attack but Kidd convinced them to let the ship go. About mid-November, they encountered a Dutch ship. It is alleged that William Moore, the *Adventure Galley* gunner, and Kidd had a dispute over whether to attack the ship or not. Kidd is said to have killed Moore by smashing a bucket over Moore's head. At the beginning of December, Kidd captured the Dutch ship the *Rouparelle.* She flew the French flag and had a French letter of marque. He renamed her the *November.*

In January, 1698 Kidd achieved his most glorious moment in

pirate annals by capturing the *Quedah Merchant,* an Armenian ship leased to the Indian government. The *Quedah Merchant* was sailing from Bengal to Surat with a rich cargo of muslins, silk, iron, sugar, saltpeter, guns and gold coins. Kidd sold some of her cargo for £10,000.

Proceeding south he captured a small Portuguese ship and returned to Saint Mary's Island off Madagascar in April 1698. The crew divided the *Quedah Merchant's* booty at the isle. Most of Kidd's crew joined up with another pirate named Culliford, and Kidd burned the *Adventure Galley.* He renamed the *Quedah Merchant* the *Adventure Prize* and left Saint Mary's in November, 1698.

The East India Company, already suffering at the hands of another English pirate, Henry Every, and now with the loss of the *Quedah Merchant* had the threat of European expulsion by the Emperor of India to contend with. The company, under pressure, compensated the merchants of the goods aboard the *Quedah Merchant,* paid bribes, and agreed to send patrols to the South Indian Sea. The Indian officials were not placated as pirate attacks continued. The British government, trying to retain their trade interests in India denounced Kidd as a pirate and omitted Kidd's name in a general pardon issued in 1698. An all-out manhunt for Kidd was ordered in November, 1698.

Kidd reached Anguilla in the Caribbean in April, 1699, in the *Adventure Prize.* Kidd learned that word had been sent out from England that he should be considered a pirate. Realizing that he would not be safe in any of the normal ports, Kidd headed for Mona Island, an uninhabited island found in the channel between Puerto Rico and Hispaniola (present day Haiti). Because Mona belonged to no one, it was a safe place to hide.

Kidd was not sure what to do. As a Scotsman, he was perhaps the last of the Scottish Templar pirates who had been dispatched by the Sinclair family of Rosslyn. Scotland and England were not on good terms at this time, and England had largely forbidden Scottish colonies and free trade across the Atlantic. The Scottish were attempting to found a colony in Panama and Kidd had heard of it. He thought of going there with his treasure. In the end, Kidd decided that he had to get to New York City, where he had influential friends and a wife that he dearly loved, and try to save himself.

The *Adventure Prize* was abandoned in the River Higuey in Hispaniola, its cargo unloaded and sold on the spot. Gold was much easier than bulky goods to transport. Kidd, now captaining the *Saint Antonio,* headed for New York City. But what happened to the *Adventure Prize* you may ask. Because its appearance was so distinctive, no one would sail it in the Caribbean. It was burned and left to sink slowly where it lay, far from its home water of the Indian Ocean.

The mood in the American colonies at this point could be characterized as one of pirate fever. Up and down the coast, everyone was on the hunt for pirates. Kidd successfully made his way to Block Island (Oyster Bay, Long Island) where he began negotiations through his contacts in New York to gain a pardon for his actions, claiming he was forced by his crew. He was induced to sail to Boston, however, where he was detained and sent to London.

Captain Kidd had entered into a fine mess indeed. His successes in the Indian Ocean nearly brought down the British East India Company, which would have been a disaster for Britain itself. In April 1698 reports reached the northern coast of India that Captain Kidd had captured a Moslem commercial ship. What's more, Kidd was carrying a bonafide document bearing the Great Seal of the King of England, and this did not look good for the burgeoning English trade in India.

Outraged governors threatened to attack East India Company warehouses and high-ranking officials moved to expel all of English traders from India. England's spectacular future in India wobbled in the balance, because if the East India Company failed, then England would have failed in India as well.

H. G. Wells in his *Outline of History* explained eloquently the odd relationship between the company and the British government:

"These successes [in obtaining the wealth of India] were not gained directly by the forces of the King of England, they were gained by the East India Trading Company, which had been originally, at the time of its incorporation under Queen Elizabeth no more than a company of sea adventurers. Step by step, they had been forced to raise troops and arm their ships. And now this trading company, with its tradition of gain, found itself dealing not merely in spices and dyes and tea and jewels, but in the revenues and territories of princes and the destinies of India. It had come to buy and sell, and it found itself achieving a tremendous piracy. There was no one to challenge its proceedings. Is it any wonder that its captains and commanders and officials, nay, even its clerks and common soldiers, came back to England loaded with spoils? Men under such circumstances, with a great and wealthy land at their mercy, could not determine what they might or might not do. It was a strange land to the English, with a strange sunlight; its brown people were a different race, outside their range of sympathy; its temples and buildings seemed to sustain fantastic standards of behaviour."

Wells concluded: "The English Parliament found itself ruling over a London trading company, which in turn was dominating an empire far greater and more populous than all the domains of the British crown."

In 1698 only about 50 British representatives, all working for the East India Trading Company, had a very tenuous hold on the Indian subcontinent. And Captain Kidd, a Scotsman, no less, was screwing it all up. Heads would roll when this was all sorted out.

In July, 1699, Kidd was captured and thrown in jail in Boston and then sent to England aboard a frigate in February, 1700, to stand trial. Once in England, Kidd became a political pawn to be used to bring down powerful men in the government. The trial started on May 8 and was completed the next day—the verdict was guilty of murder and multiple piracies.

Captain William Kidd was hanged on May 23, 1701, but not easily. The first rope put around this neck broke so he had to be strung up a second time. After his death, his body was covered in tar, bound in chains, his head covered with a metal harness and hung at Tilbury Point as a warning to all those passing. The body remained there until it was totally rotted away. Captain Kidd would never sail again, but a legend grew up around his treasure.

§§§

Captain Kidd's Treasure

The disappearance of most of his booty gave rise to legends about Kidd and his buried treasure. The only treasure actually recovered was found on Gardiners Island, off Long Island, in 1699. Literary treatments of the Kidd legend include Edgar Allan Poe's *The Gold Bug* (1843).

Kidd's treasure is perhaps the most famous of all pirate treasures. Even the Oak Island Money Pit was rumored to contain Kidd's fabulous treasure of gold bars and coins.

How much gold did he actually have? What happened to it? Did he bury some part of it while he lay at anchor at Block Island? Could he have gone up the Connecticut River, portaged around the falls he encountered, and found a good hiding place on Clark's Island? We can't know for sure, but present day maps of the Connecticut River label the island as Kidd's Island.

Clark's Island, which lies in the Connecticut River in Northfield, Massachusetts, just off the upper end of Pine Meadow, has a legend attached to it. According to Temple and Sheldon (1875), the story goes this way: Captain Kidd and his men ascended the Connecticut River searching for a place to bury a treasure of gold, somewhere secluded but distinctive. They buried the chest of gold and drew lots to see which of their number would be killed so that his body could be left on top of the chest to protect it from all treasure hunters. Over the years a legend grew up around the treasure—the gold could be dug up only by three people at midnight when the full moon was directly overhead. They must form a triangle around the exact spot and work in absolute silence, words would break the charm!

In the early 19th century, Abner Field and two of his friends attempted to find the treasure by following the directions exactly. At midnight, under a full moon shining directly on them, they sweated and dug, silently. Shovelful by shovelful, they dug deeper and deeper. The sweat poured off their bodies even in the chill night air. The mosquitoes swarmed around, biting, but the three men were afraid to kill them for fear the sound would break the charm. Any amount of discomfort could be tolerated in order to find the buried treasure chest.

Suddenly there was the echoing sound of crowbar striking against iron. Just as the men saw a corner of the chest emerge from the dirt, someone exclaimed, "You've hit it!" and the trio of treasure hunters watched in consternation as the chest immediately began to sink out of reach...

And so, like so many tales of treasure and treasure maps, we know it is there, but it is just out of reach!

Sir Francis Drake, whose name means "dragon."

Left: The typical Jolly Roger. Right: The flag of Bartholomew Roberts.

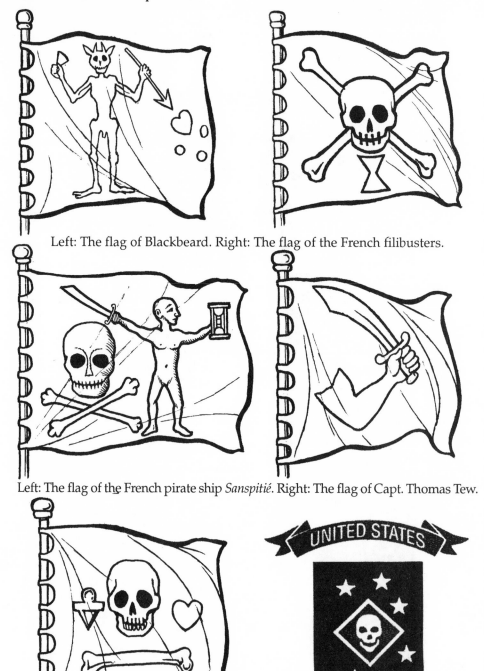

Left: The flag of Blackbeard. Right: The flag of the French filibusters.

Left: The flag of the French pirate ship *Sanspitié*. Right: The flag of Capt. Thomas Tew.

Left: The flag of Stede Bonnet. Right: The flag of the U.S. Marine Raiders.

Rackham's distinctive flag.

"Calico" Jack Rackham.

Pirate fevor gripped the streets of New York during the time of Captain Kidd.

The trial of Captain Kidd.

Captain Kidd at the gallows.

Captain Kidd hung on the Thames as an example to other pirates.

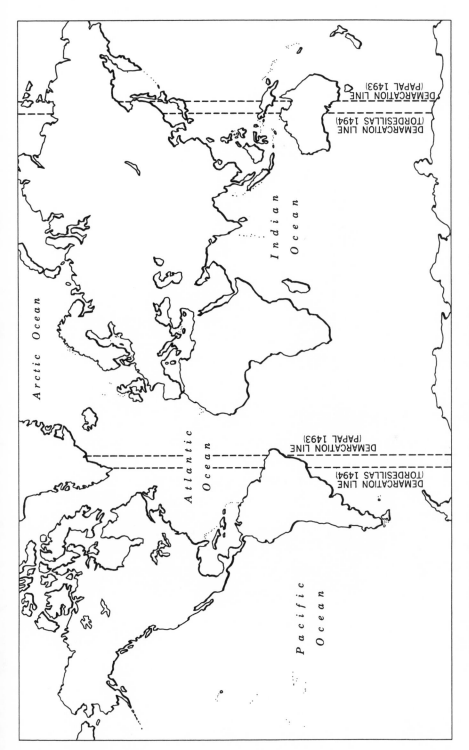

The Papal Demarcation Lines of 1493 and 1494, dividing the world between Spain and Portugal.

George Washington in his Masonic gear.

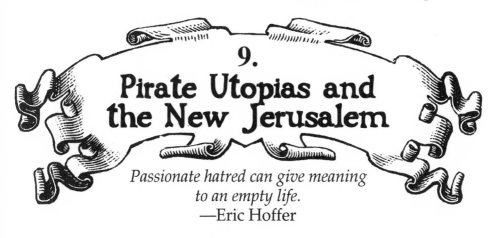

9.
Pirate Utopias and the New Jerusalem

*Passionate hatred can give meaning
to an empty life.*
—Eric Hoffer

"I have not yet begun to fight!"
—John Paul Jones

The Templar Mission of a New Jerusalem

The legacy left by the Knights Templar on the modern world should not be underestimated. Their pirate fleet roamed the North Seas for centuries, preying on ships that were loyal to the Vatican. As the new lands of Nova Scotia and America began to be settled, the Knights Templar and their fleet were there as well. In their new guise as the Scottish Rite of Masons, they were instrumental in creating the United States of America, if nothing else.

The Templars began their modern transformation at Rosslyn castle according to Tim Wallace-Murphy in his book, *The Templar Legacy & the Masonic Inheritance within Rosslyn Chapel*.[13] According to Wallace-Murphy, the builder of Rosslyn Chapel, William St. Clair, was the last Sinclair 'Jarl' (Earl) of Orkney, who lived in the middle of the fifteenth century. After Earl William, the 'Jarldom of Orkney' passed from the family to the Scottish crown as part of the dowry of Margaret of Denmark on her marriage to King James III of Scotland. William was not only the grandson of Prince Henry Sinclair and the last Jarl of the Orkneys, he also had the somewhat peculiar title of Knight of the Cockle and the Golden Fleece. According to Wallace-Murphy, Sir William St. Clair was a member of a secret group that preserved important knowledge concerning the Holy Grail, the Holy Blood of the Merovingian kings, and the destiny of the new continent across the Atlantic.

In *The Temple and the Lodge*,[3] Baigent and Leigh say that, "...by the Middle Ages, the architect or builder of Solomon's Temple had already become significant to the guilds of 'operative' stonemasons. In 1410, a manuscript connected with one such guild mentions the 'king's son of Tyre' [Hiram], and associates him with an ancient science said to have survived the Flood and been transmitted by Pythagoras and Hermes. A second, admittedly later, manuscript, dating from 1583, cites Hiram and describes him as both the son of the King of Tyre and a 'Master.' These written records bear testimony to what must surely have been a widespread and much older tradition."[3]

Clearly, the Knights Templar saw themselves as the inheritors of ancient knowledge that went back to Atlantis. They struggled for hundreds of years against the Vatican and the reign of terror known as the Inquisition. To the Templars, the true church, one that taught mysticism, reincarnation and good works, was being suppressed by a dark power that called itself the one true faith. Oppression of these other faiths was accomplished through the familiar devices of torture, terror and extermination.

Henry Sinclair of Orkney had risked all to make his voyages across the North Atlantic. Had he taken the Holy Grail and possibly even the Ark of the Covenant to America? Had these sacred relics helped spur on the creation of the United States, a land which Masonic founding fathers like George Washington, Thomas Jefferson and Benjamin Franklin were to create partially on the Templar ideals of religious freedom and a free-trade banking system?

According to Templar historians like Michael Baigent, Richard Leigh, Andrew Sinclair and Tim Wallace-Murphy, the Knights Templar had helped create an independent Scotland, then a "New Scotland" and finally an independent United States.

Mystic Christian themes such as the Holy Grail and the New Jerusalem run deep in our study of the lost Templar fleet and its successors. We have already discussed Sir Francis Bacon and his unfinished utopian romance *The New Atlantis,* published around the year 1600, and raised the idea that this book was reflecting the desires of the remaining Templar Order to use the fleet based in Scotland to create a utopian New Jerusalem in the lands across the North Atlantic.

A number of authors have claimed that the Templars did exactly that, and that Montreal was founded to be the New Jerusalem of the "New Atlantis." Canadian author Michael Bradley maintains in his book *Holy Grail Across the Atlantic* that the French explorer and founder of the Quebec colony Samuel de Champlain (1567-1635) was a secret agent for the Grail Dynasty and the Grail was moved to Montreal just before Nova Scotia was attacked by the British Admiral Sedgewick in 1654. A mysterious secret society called the Compagnie du Saint-Sacrement carried the Grail to Montreal.[9]

Another Canadian author, Francine Bernier says that evidence exists that the Templars were attempting to make the city of Montreal the New Jerusalem. In her book *The Templar Legacy: Montreal, the New Jerusalem,* Bernier maintains that there is a secret history to the founding of the city. She claims that the island of Montreal was designed in the 17th century as the New Jerusalem of the Christian world to become the new headquarters of a group of mystics who wanted to live as the flawless Primitive Church of Jesus. The people behind the scene in turning this dream into reality were the Société de Notre-Dame, half of whose members were in the elusive Compagnie du Saint-Sacrement. They took no formal vows and formed the interior elitist and invisible "heart of the church" following a "Johannite" doctrine of the Essene tradition, where men and women were considered equal apostles.[33]

Bernier claims that these new Christian Crusaders made sure their noble goal would be continued abroad, in "Nouvelle France," with Montréal as the New Jerusalem promised in the Bible. The book reveals the links between Montreal and: (1) John the Baptist as patron saint; (2) Melchizedek, the first king-priest and a father figure to the Templars and the Essenes; (3) Stella Maris, the Star of the Sea from Mount Carmel; (4) the Phrygian goddess Cybele as the androgynous Mother of the Church; (5) St. Blaise, the Armenian healer or "Therapeut"—the patron saint of the stonemasons and a major figure to the Benedictine Order and the Templars; (6) the presence of two Black Virgins, particularly one from Montaigu; (7) a Sulpician chapel which is based on the theme of the Temple of Solomon.[33]

In light of much of what we have now learned about the lost Templar fleet, the Holy Blood-Holy Grail, and the Sinclair family of Scotland, it seems reasonable to conclude that Montreal may well have been meant to be the New Jerusalem. The eastern seaboard of Canada and New England, plus the lands that could be reached by the St. Lawrence river were areas which the Templars were targeting for a massive resettlement campaign. This plan would take hundreds of years to fully achieve, but the Templars were clearly thinking ahead—very far ahead. They were engineering the future in the way that they wanted the future to go. Their goals were to create new countries that would be free from religious oppression and false doctrine of the "divine right of kings."

If the early goals of the Templar fleet were to establish outposts on Oak Island and other areas of Nova Scotia, and then create Templar cities such as Montreal, their next goal would be to create an independant country that was free from European Kings and the influence of the Vatican. This would mean a revolution—an American revolution.

§§§

The Knights Templar and the American Revolution

It is estimated that over 80% of the founding fathers of America, the signers of the Declaration of Indepencence, were Scottish Free Rite Masons, and therefore descendants of the Knights Templar and the societies that preceded them. The American Revolution was not so much a battle against the King of England and Scotland—King George—but a battle to create a Templar/Masonic estate of the "Builders" who would be free of any hereditary monarchy or other corruptible system of government.

Even servants of the King, initiates of the Masonic Order, were working against him and the royalty, because they had been instructed to do so. Though it was not well known, an inner core of the Scottish Order was coordinating, as best they could, the creation of the United States and the defeat of the British Royal Crown. The Knights Templar/Masons were behind the scenes in the creation of the United States, seeing to it that certain bridges were left unguarded and that Washington and his troops could escape when they were surrounded.[2]

Though King George back in Britain wasn't let in on these plans to create an independant United States, separate from his royal power, the ancient societies nonetheless allowed him to wage war on the colonies, knowing he would be defeated. In modern Masonic mythology, two brothers, both Masons, were commanders in the British Red Coat forces against the revolutionaries, but because they were Masons, allowed Washington to escape from time to time to regroup his forces and continue the battle.[2]

Among the efforts of General Washington to elude the British forces, apparently with the help of the Masons, was the use of the "Culver Spy Ring." In June of 1778, British General Sir Henry Clinton's forces occupied the majority of New York City, while General George Washington's troops were scattered throughout the area. The war was only in its second year, but things were going badly for the colonies and Washington desperately needed intelligence regarding the movements of Clinton's forces. Long Island, like New York City, was occupied by the British when Washington's people recruited Long Island resident (and Mason) Abraham Woodhull to do intelligence work for the rebelling colonies. His code name, and alias, became Samuel Culver.

Woodhull's first move was to recruit a New York City businessman named Robert Townsend, also a Mason. In turn, Townsend recruited agents of his own: his wife and his brother-in-law. Together they formed the Culver Spy Ring, which would provide important information regarding Clinton's troop movements for Washington's forces. Washington instructed them to write the information on the pages of common pocket books of the time. Then these books would be delivered to Washington.

Washington didn't know the identities of the spies, but knew they were Masons. He wanted them to mix with the British and Tories, to visit their coffeehouses and all their public places—and they did. Townsend, for one, wrote a gossip column for a Tory newspaper which gave him increased access to the enemy. One contact he cultivated was Major John Andre who would later turn Benedict Arnold over to the British side. Townsend would take his reports to a tavern keeper named Austin Roe. Roe, in turn, would forward them on to Woodhull's farm. The information then was sent through one of Woodhull's couriers to Washington.

The messages gave Washington the day-by-day movements of Clinton's troops. This way Washington could stay one step ahead of the British forces since a direct confrontation with Clinton's forces could wipe out the revolutions' ragged army. The information turned out to be priceless. Washington knew exactly what the British were doing and what they were planning. An example of how this type information could be used can be illustrated by the arrival of French troops in 1780. By that time, the many efforts by Benjamin Franklin to get help for the American cause had come to fruition. When French troops arrived in Newport, Rhode Island, Clinton's troops prepared to attack the French before they could join up with Washington's army.

The Culver Spy Ring went into action. They saw the British preparing for an invasion of Newport and reported this to Washington. Washington, in turn, decided to divert the attention of the British by preparing an invasion of Manhattan. This news got back to the British, because they too had their spies, and Clinton decided to call off the attack. This flurry of disinformation was a major turning point in the war. The British had been fooled and the forces of France and Washington's army did meet. The American forces and the French were able to defeat the British and the colonies won their independence. Throughout the revolution the colonies were able to gather information that was crucial to an American victory.

With the creation of United States of America, a new age had begun in politics, taxation, science, religious freedom, and multicultural acceptance. America became the melting-pot of the world, where the best from all societies, races, creeds, and philosophies came in search of the golden age that their ancient texts had always predicted. The United States would become strong because it welcomed all people, all ideas, all religions (except the intolerant and violent), and most importantly, all people with the burning desire for a better life and an idealistic and spiritual lifestyle. With such innocence and naiveté, the United States of America entered into the history of the world.

§§§

The Masonic "Pirates" of the Revolution

As the American Revolution is now hundreds of years behind us, it has been analyzed and written about by historians around the world. This amazing event, so successful and encouraging to oppressed peoples everywhere, was to set the stage for other revolutions to follow—many not so successful or well-conceived.

The British crown had many advantages, but the Americans were not without their strong points. For instance, a vast reservoir of manpower could be drawn upon. For the most part, men preferred short-term enlistments, and many who served came out for only a few weeks or months—but they did serve. The best estimates are that over 200,000 participated on the patriot side. General Washington was often short of shoes and powder, but rarely were he and other commanders without men when they needed them most, although at times American leaders had to take into the army slaves, pardoned criminals, British deserters, and prisoners of war. Moreover, Americans owned guns, and they knew how to use them.

If the Continental Army won few fixed battles, it normally fought reasonably well and it extracted a heavy toll on the enemy, who usually could not easily obtain reinforcements. Although only Washington and Maj. Gen. Nathanael Greene were outstanding commanders, many others were steady and reliable, including Henry Knox, Benjamin Lincoln, Anthony Wayne, Daniel Morgan, Baron von Steuben, the Marquis de Lafayette, and Benedict Arnold, before he defected to the enemy in 1780. All of these men were Masons, including Benedict Arnold.

The Americans had tremendous advantages on land. They had no obvious or central point to attack. It was discouraging to seize at one time or another every American urban center and yet have nothing more to show for it than the mere possession of territory, since the Americans had no single vital

strategic center. The British were waging war 3,000 miles from Europe against an armed population spread over hundreds of miles, from the Atlantic to the Mississippi, from Maine to Georgia. The land was forested, ravined, swampy, and interlaced by myriad streams and rivers.

The only real advantage that Britain had over the Continental forces was that of overwhelming naval power. It has been said that the country with the largest navy rules the world, and during the American Revolution, this country was Britain. To overcome the superior naval force of the British, the Americans would turn to pirates. The lost fleet of the Knights Templar would now be used against the King of England to build a new world, much as the Templars had helped sustain an independent Scotland against the English in the past.

To the British it seemed to take forever—6 to 12 weeks—for word of campaign strategies to pass from London to commanders in the field, for provisions to arrive, and for naval squadrons to appear in time for cooperation with land forces. The scope of the contest also reduced the Royal Navy's effectiveness in blockading the long American coastline.

The New England coast alone was a vast territory for the British navy to patrol. Stores could be landed at too many rivers, bays, and inlets. Nor could the British employ their fast frigates and formidable ships of the line (battleships of the 18th century) against an American fleet. Such a fleet simply did not exist. The fledgling republic had no official navy.

The revolutionary navy that did go into battle against the powerful British admiralty during the war was a Masonic-Templar-Pirate force that chose to take on British ships in a one-on-one style of naval broadsides.

Patriots took to the sea in single ships, either privateers or vessels commissioned by Congress. It should be noted here that since the Continental Congress was not a recognized government, even commissioned ships were often viewed as pirates. Consequently, the British-American naval war can be told largely as a story of individual ship duels between pirates and British man-o-wars.

The triumph of John Paul Jones in the *Bonhomme Richard* over *H.M.S. Serapis* in the North Sea in 1779 was the most famous of these encounters. John Paul Jones was to go down in history as the father of the American navy, and a hero of the Revolution—and he was a pirate, a privateer, a freebooter, and a Scottish Free Rite Mason.

Born on July 6, 1747, to a Scottish gardener, his full name was originally John Paul. At the age of 12 he entered the British merchant marine. Sailing aboard merchantmen and slavers, he received his first command in 1769.

In 1773, however, in an unfortunate incident, he killed a mutinous crewman. On the Caribbean island of Tobago, where his ship *Betsy* ended her outward voyage, John Paul decided to invest money in return cargo rather than pay his crew for their shore leave. One sailor, known as "the ringleader," attempted to go ashore without leave. John Paul drew his sword on the man to enforce his orders, but the man set on his captain with a bludgeon. In response to the attack John Paul ran him through with his sword. He immediately went ashore to give himself up, but the death of the ringleader had so stirred up local sentiment that John Paul's friends prevailed upon him to escape to Virginia at once. From that point on, the British considered him to be a pirate. A fugitive from British justice, he attempted to conceal his identity by taking on the surname of Jones.

At the outbreak of war with Britain in 1775, John Paul Jones went to Philadelphia, and, with the help of two friendly members of the Continental Congress, obtained a lieutenant's commission in the Continental Navy.

On 3 December 1775, as first lieutenant of the *Alfred*, he hoisted the Grand Union flag for the first time on a Continental warship. In February 1776 John Paul Jones participated in the attack on Nassau, New Providence Island. Jones was appointed to command the sloop *Providence* on 10 May 1776; his commission as Captain in the Continental Navy was dated 8 August 1776. In his first adventure as captain, he destroyed the British fisheries in Nova Scotia and captured 16 British prize ships! The 12-gun sloop departed for the Delaware Capes on 21 August. Within a week she had captured the whaling brigantine *Britannia*.

Near Bermuda, she fell upon a convoy escorted by the 28-gun frigate *Solebay*.

In a thrilling chase lasting ten hours, Jones saved *Providence* from the larger warship by an act of superior seamanship. By 22 September he had captured three British merchant vessels. While anchored he burnt an English fishing schooner, sank another, and made prize of a third.

Clearly, Jones was on a major roll! He would later declare that his best crew had been the one aboard the *Providence*. He had received sound financial rewards from the prizes, making this venture the all-around most enjoyable of his career.

In November 1777, John Paul Jones sailed for France in the *Ranger*, carrying word of Burgoyne's surrender at Saratoga. Admiral La Motte-Picquet returned Jones' salute at Quiberon Bay on 14 February 1778, marking the first time the Stars and Stripes were recognized by a foreign power. During that spring he terrorized the coastal population of Scotland and England by making daring raids ashore. *Ranger* later captured the British sloop of war *Drake* off the coast of Ireland on 24 April, and pillaged the British coast.

His reputation in Paris greatly enhanced, Jones received from the French government a converted French merchantman, *Duras*, which he renamed *Bonhomme Richard* (Poor Richard) in honor of Benjamin Franklin. Franklin was a fellow Mason, and the influence of French Masons in the creation of the United States is considerable.

Setting sail from France at the head of a small squadron on Aug. 14, 1779, Jones captured 17 merchantmen off the British coast and on September 23 fell in with a convoy of British merchant vessels escorted by *H.M.S. Serapis* and *Countess of Scarborough* off Flamborough Head, Yorkshire. One of the most exciting naval battles of all time was about to begin. *Serapis* was a superior ship compared to *Richard*. She was faster, more nimble and carried a far greater number of eighteen pounders. The two ships fired simultaneously. At the first or second salvo, two of Jones' eighteen pounders burst, killing many gunners and ruining the entire battery, as well as blowing up the deck above. After exchanging two or three broadsides, and attempting to rake the *Serapis'* bow and stern, the commodore decided that he must board and grapple, a gun-to-gun duel seeming futile.

Serapis' Captain Pearson repulsed the boarders, and attempted to cross *Richard's* bow to rake her. During this stage of the bloody and desperate battle, Pearson, seeing the shambles on board *Bonhomme Richard*, asked if the American ship had struck (i.e., was incapacitated).

Although his smaller vessel was on fire and sinking, Jones rejected the British demand for surrender: "I have not yet begun to fight," he declared. His immortal reply then served as a rallying cry to the crew.

The two ships grappled and Jones relied on his marines to clear the enemy's deck of men. With the muzzles of their guns touching, the two warships continued to fire into each other's insides.

Then, to Jones' great amazement, the *Alliance,* under the Frenchman Pierre Landais, fired three broadsides into *Richard.* In this heated battle between pirates and British lords (as virtually all English naval officers were) had his French ally turned against him? Landais later stated that he wanted to help *Serapis* sink *Richard*, then capture the British frigate.

Even though his ship had begun to sink, Jones determined he would not strike his colors. The battle continued on, with the cannons of both ships firing point blank into the other. He used his remaining guns to weaken *Serapis'* main mast. It began to tremble, and more than three hours after the bloody battle began, Pearson lost his nerve and decided to strike his colors. Jones, incredibly, had won the day!

When the battered *Bonhomme Richard* sank on 25 September, Jones was forced to transfer to *Serapis*. It was a major victory for the fledging American navy, and news of the victory dampened spirits in England and gave great hope to the Masonic revolutionaries. For his victory, Congress passed a resolution thanking Jones, and Louis XVI presented him with a sword.

Although hailed as a hero in both Paris and Philadelphia, Jones encountered such stiff political rivalry at home that he never again held a major American command at sea. In 1788, Russian Empress Catherine II (The Great) appointed him rear admiral in the Russian navy. He took a leading part in the Black Sea campaign against the Ottoman Turks. Jealousy and political intrigue among his Russian rivals prevented him from receiving proper credit for his successes and resulted in his discharge.

In 1790 he retired and went to live in Paris. In 1792 Jones was appointed U.S. Consul to Algiers, but on July 18 of that year he died before the commission arrived. He was buried in Paris, but in 1905 his remains were removed from his long-forgotten grave and brought to the United States where, in 1913, they were finally interred in the U.S. Naval Academy Chapel at Annapolis, Maryland. This Scottish-Masonic pirate is considered the founder of the American navy!

§§§

Legacy of John Paul Jones

Most general histories of Jones and his involvement with the U.S. Navy focus on his seamanship and courage when in danger, but not so much on his strong character. The honorable portrayals of Jones seen by the American and French audiences are in contrast to images of piracy presented by the British.

Rudyard Kipling, for example, referred to the "exploits" of Jones, "an American Pirate." Sir Winston Churchill called him a "privateer." But Theodore Roosevelt mentions him as a "daring corsair."

Jones, of course, neither held a privateering commission nor was engaged in piracy by the strict definition of the word, since he had a commission from the Continental Congress. But when Jones was raiding British ships at the start of the war, they probably thought of him as a privateer and a scurvy dog, not as the Admiral of some respectable navy. It is amusing to note that the British eventually came to the the realization that Jones was not technically a "pirate" which prompted someone on one occasion to cross out the British Library catalogue entry for John Paul Jones as "the Pirate," and substitute "Admiral in the Russian Navy."

Critics of John Paul Jones have said that he indulged in questionable behavior, since his popularity with women led him to having many lovers, and moreover, he failed to be a good team player, spurning the naval efforts of others as inadequate compared with his own brilliant accomplishments. His arrogance is probably what kept him from advancing in the American and

Russian navies, but his accomplishments cannot be denied.

John Paul Jones not only had a brilliant naval career, he also wrote in detail throughout his life to promote professional naval standards, training and protocol. For generations, midshipmen have been required to memorize his dicta outlining the appropriate qualifications and duties of a naval officer.

"None other than a Gentleman, as well as a seaman, both in theory and practice is qualified to support the character of a Commissioned Officer in the Navy, nor is any man fit to command a Ship of War who is not also capable of communicating his Ideas on Paper in Language that becomes his Rank." —John Paul Jones to Marine Committee, 21 January 1777

"As you know that the Credit of the Service depends not only on dealing fairly with the men Employed in it, but on their belief that they are and will be fairly dealt with." —John Paul Jones to Joseph Hewes, 30 October 1777

§§§

Washington's French Masonic Counterpart: General Lafayette
Another important Masonic figure in the American Revolution was the French general the marquis de Lafayette. Called the hero of two worlds, he played a prominent role in both the American Revolution and the French Revolution. Born on Sept. 6, 1757, to a noble family in the Auvergne, he defied the French authorities in 1777 by crossing the Atlantic to offer his services to the Continental Congress at Philadelphia. Lafayette was a known Mason, and even though he was never officially an American citizen, he has an important chapter in the book *Masonic Membership of the Founding Fathers*.[78]

He was a friend of George Washington, who became his model, and served under him at the Battle of Brandywine and at Valley Forge. In 1779 he went to France to expedite the dispatch of a French army, but he returned to distinguish himself again at the battle of Yorktown in 1781. Brave in battle and staunch in adversity, Lafayette won enduring popularity in America, and his fame did much to make liberal ideals acceptable in Europe.

As discontent in France mounted under the rule of Louis XVI, Lafayette advocated the establishment of a representational monarchy (something like that which existed in England), and represented the nobility of Auvergne in the States General in 1789. He became a deputy and proposed a model Declaration of Rights. Elected commander of the National Guard on July 15, 1789, he appeared gallantly with his troops at the Festival of Federation on July 14, 1790, to celebrate the apparent coming of age of a free and united community. However, Lafayette proved unable to fulfill the promise of his youth. Although he was truly inspired by the ideals and success of the American Revolution, and he had enormous potential power as a mediator, he had neither a realistic policy of his own nor the flexibility to support the more practical comte de Mirabeau. Now a man in the middle, he was despised by the court as a renegade aristocrat whose bourgeois army was unable to protect the royal family, and he was also hated by the populace for trying to suppress disorder, especially after he fired on a crowd in Paris in July 1791.

In 1792, Lafayette became an army commander when war was declared against Austria, but he was ousted with the rise of the Jacobites. He was caught by the Austrians, and turned over to the Prussians who promptly imprisoned him as a dangerous revolutionary. Released in 1797 at Napoleon Bonaparte's insistence, Lafayette was allowed to return to France in 1799. In 1815 he was one of those who demanded Napoleon I's abdication. A republican at heart, he could never fully support the overthrow of the monarchy for an empire.

In 1824, Lafayette made a triumphant tour of the United States. By then his Parisian estate, La Grange, was a place of pilgrimage for liberals throughout the world. When the July Revolution of 1830 occurred, he was again called on to command the French National Guard, to try to identify the monarchy of Louis Philippe with the ideals of 1789. He died in Paris on May 20, 1834; his name continues to signify freedom, in a pirate kind of way.

§§§

Privateers and the War of 1812

The newly created United States of America was quickly to have to go back into naval battle, fighting an odd combination of foes—a bunch of North African pirates and the British. As the nascent nation struggled to gain international recognition and establish sea trade, obstacles arose stemming from both the continents of Africa and Europe. First, let us look at the War of 1812, in which the Americans were again helped by both privateers and Masonic sea captains.

The War of 1812 was fought between the United States and Great Britain from June 1812 to the spring of 1815, although the peace treaty ending the war was signed in Europe in December 1814. The main land-force fighting of the war occurred along the Canadian border, in the Chesapeake Bay region, and along the Gulf of Mexico; extensive action also took place at sea, and pirates fighting for the Americans were a major help.

From the end of the American Revolution in 1783, the United States had been irritated by the failure of the British to withdraw from American territory along the Great Lakes, their backing of the Indians on America's frontiers, and their unwillingness to sign commercial agreements favorable to the United States. American resentment grew during the French Revolutionary Wars (1792-1802) and the Napoleonic Wars (1803-15), in which Britain and France were the main combatants.

In time, France came to dominate much of the continent of Europe, while Britain remained supreme on the seas. The two powers also fought each other commercially: Britain attempted to blockade the continent of Europe, and France tried to prevent the sale of British goods in French possessions.

The British Orders in Council of 1807 tried to channel all neutral trade to continental Europe through Great Britain, and France's Berlin and Milan decrees of 1806 and 1807 declared Britain in a state of blockade and condemned neutral shipping that obeyed British regulations. The United States believed its rights on the seas as a neutral were being violated by both nations, but British maritime policies were resented more because Britain dominated the seas. Also, the British claimed the right to take from American merchant ships any British sailors who were serving on them.

Frequently, they also took Americans. This practice of impressment became a major grievance.

The United States at first attempted to change the policies of the European powers by economic means. In 1807, after the British ship *Leopard* fired on the American frigate *Chesapeake*, President Thomas Jefferson urged and Congress passed an Embargo Act banning all American ships from foreign trade. The embargo failed to change British and French policies but devastated New England shipping. This led to a natural increase in illegal shipping.

Failing in peaceful efforts and facing an economic depression, some Americans began to argue for a declaration of war to redeem the national honor. The Congress that was elected in 1810 and met in November 1811 included a group known as the War Hawks who demanded war against Great Britain.

They argued that American honor could be saved and British policies changed by an invasion of Canada. The Federalist party, representing New England shippers who foresaw the ruination of their trade, opposed war. Napoleon's announcement in 1810 of the revocation of his decrees was followed by British refusals to repeal their orders, and pressures for war increased. On June 18, 1812, President James Madison signed a declaration of war that Congress—with substantial opposition—passed at his request. Unknown to Americans, Britain had finally, two days earlier, announced that it would revoke its orders.

U.S. forces were not ready for war, and American hopes of conquering Canada collapsed in the campaigns of 1812 and 1813. The initial plan called for a three-pronged offensive: from Lake Champlain to Montreal; across the Niagara frontier; and into Upper Canada from Detroit. The attacks were uncoordinated, however, and all failed. In the West, Gen. William Hull surrendered Detroit to the British in August 1812; on the Niagara front, American troops lost the Battle of Queenston Heights in October; and along Lake Champlain the American forces withdrew in late November without seriously engaging the enemy.

American frigates won a series of single-ship engagements with British frigates, and American privateers continually harried British shipping. The captains and crew of the frigates *Constitution* and *United States* became renowned throughout America. Meanwhile, the British gradually tightened a blockade around America's coasts, ruining American trade, threatening American finances, and exposing the entire coastline to British attack.

American attempts to invade Canada in 1813 were again mostly unsuccessful. There was a standoff at Niagara, and an elaborate attempt to attack Montreal by a combined operation involving one force advancing along Lake Champlain and another sailing down the Saint Lawrence River from Lake Ontario failed at the end of the year.

The Americans won control of the Detroit frontier region when the ships of Oliver Hazard Perry destroyed the British fleet on Lake Erie (Sept. 10, 1813). This victory forced the British land forces to retreat eastward from the Detroit region, and on Oct. 5, 1813, they were overtaken and defeated at the battle of the Thames (Moraviantown) by an American army under the command of Gen. William Henry Harrison. In this battle the great Shawnee chief Tecumseh, who had harassed the northwestern frontier since 1811, was killed while fighting on the British side.

But, in 1814 the United States faced complete defeat, because the British, having defeated Napoleon in Europe, began to transfer large numbers of ships and experienced troops to America. The British planned to attack the United States in three main areas: in New York along Lake Champlain and the Hudson River in order to sever New England from the union; at New Orleans to block the Mississippi; and in Chesapeake Bay as a diversionary maneuver.

The British appeared near success in the late summer of 1814. American resistance to the diversionary attack in Chesapeake Bay was so weak that the British, after winning the Battle of Bladensburg on August 24, marched into Washington, D.C., and burned most of the public buildings. President Madison had to flee into the countryside.

The British then turned to attack Baltimore but met stiffer resistance and were forced to retire after the American defense of Fort McHenry (which inspired Francis Scott Key to write the words of the "Star-Spangled Banner").

In the north, about 10,000 British veterans advanced into the

United States from Montreal. Only a weak American force stood between them and New York City, but on Sept. 11, 1814, American Capt. Thomas Macdonough won the naval battle of Lake Champlain, destroying the British fleet. Fearing the possibility of a severed line of communications, the British army retreated into Canada.

When news of the failure of the attack along Lake Champlain reached British peace negotiators at Ghent, in the Low Countries, they decided to forego territorial demands. The United States, although originally hoping that Britain would recognize American neutral rights, was happy to end the war without major losses. The Treaty of Ghent, signed by both powers on Dec. 24, 1814, supported, in essence, the conditions in existence at the war's onset!

Because it was impossible to communicate quickly across the Atlantic, the British attack on New Orleans went ahead as planned, even though the war had officially ended, and isolated naval actions continued for a few months.

Pivotal in this battle was a certain French pirate named Jean Lafitte (c.1780-c.1826), a Louisiana privateer and smuggler. About 1810 he and his men settled in the area of Barataria Bay, near New Orleans, and preyed on Spanish ships in the Gulf of Mexico. In 1814 the British attempted to buy Lafitte's aid in attacking New Orleans. Instead he passed their plans on to the Americans, and helped Andrew Jackson defend the city in January 1815. Lafitte later returned to privateering.

Andrew Jackson's decisive victory at New Orleans caused the British to suffer more than 2,000 casualties; the Americans, fewer than 100. Because of the lag in communications, the signing of the peace treaty was popularly linked with Jackson's victory at New Orleans. This convinced many Americans that the war had ended in triumph, and a surging nationalism swept the country in the postwar years.

§§§

The Barbary States and Pirate Utopias

In the years just prior to the War of 1812, a curious bit of American and North African history took place: the newly created United States of America and the Pirate Utopias of North Africa went to war.

The Barbary States is a former name for the coastal region of North Africa extending from the Atlantic Ocean to Egypt, and comprising the present states of Morocco, Algeria, Tunisia, and Libya. The name is derived from the Berbers, the oldest known inhabitants of the region.

In ancient times, parts of the region were colonized by the Phoenicians, who controlled the eastern Mediterranean but later created a powerful city named Carthage to the west, in what is today in Tunisia. North African piracy is said to date from the time of the Phoenicians. The Romans fought three wars with the Phoenician city-state of Carthage, and by the middle of the 1st century AD the area was in Roman hands. It remained Roman, ultimately allowing them to dominate the Mediterranean, until the Vandals invaded in the 5th century, was briefly recovered (533-34) by the Eastern Roman (Byzantine) Empire, and was overrun by the Muslim Arabs in the 7th century.

In the 16th century the Arab principalities of North Africa came under the nominal rule of the Ottoman Empire. They were in fact conquered for Turkey by a corsair, or pirate, known as Khair ad-Din Barbarossa (c.1483-1546) to prevent their falling to Christian Spain. Barbarossa and his brother, Koruk or Aruj (c.1474-1518), were the most famous of a large band of corsairs active in the region. Under power from the Ottoman Empire, these corsairs, like the Carthaginian navy before them, harassed all the Roman Catholic ships that they could find. The Barbarossa brothers siezed Algiers from the Spanish in 1518 and put Algeria under Turkish suzerainty. As admiral of the Turkish-Barbary fleet, Khair ad-Din twice defeated the Genoese Admiral Andrea Doria, once at Tunis and once at Algiers. The brothers were much feared along the shipping lanes around Malta and Sicily—the area where the Jolly Roger had apparently originated. At the mere mention of the

Barbarossa name in Europe, "...men swore, and women crossed themselves."[32]

The Barbary States became a base for piracy against European shipping in the Mediterranean. The booty—and tribute paid to gain immunity from attacks—were the chief source of revenue for the local rulers. Europe, and even America, were starting to stabilize after hundreds of years of religious wars, piracy, revolutions and technological advances. The pirates and their "barbaric" acts—often exaggerated—were just not to be tolerated anymore.

But how did these pirate states spring up? Beat writer Peter Lamborn Wilson defends these pirate states as utopian communities in his book *Pirate Utopias*.[32] Wilson has written on sufism and the Assassins, and in this book, seeks to show that pirate utopias and other "insurrection communities," as he calls them, have their origin in pirate culture.

In the book he focuses on the corsairs' most impressive accomplishment, the independent "Pirate Republic of Salé." It deals with European sailors who convert to Islam while in North Africa, most often "pirates," who then were known as "renegadoes" and eventually, "corsairs."

The Pirate Republic of Salé, according to Wilson, was the most important of all pirate utopias, one that was actually recognized by European powers for a brief time. Salé is a small port just north of Rabat in present-day Morocco, on the Atlantic side of the Straits of Gibralter. From Salé the corsairs, known also as Salee Rovers, terrorized shipping that was flowing between the Atlantic and the Mediterranean.

Wilson describes the end of this "pirate utopia" thusly:

> The Bou Regreg Republic may have lost some autonomy under the regime of the Dala'iyya, but perhaps gained—at last—some peace and balance under the nominal *saltanat* of the Sufi order. In any case, the last two decades of the Triple Republic were its most golden, at least in terms of piracy. Freed at last of internecine strife, all three city-states could turn all their hostility outward-in the corsair holy war. Moreover, if the corsair republics in their purest form (1614-1640) were unique as political entities, one can only use a pleo-

nasm like *"more* unique" to describe the condominium-régime of corsairs and Sufis, which lasted from 1640 to 1660. It boggles the imagination-and indeed it was too good to last long. The hand of the Dala'iyya and its chief in Salé—Sidi Abdullah the "prince of Salé"—came to feel heavier and heavier to the Andalusians and pirates. They began to look for some means to restore their pristine state of total independence, which by now had come to take on all the aura of an ancient and revered tradition.

Meanwhile... a disciple of the martyred marabout al-Ayyashi, an Arab from Larache (and therefore an enemy of the Dala'iyya Berbers, those "shirtless animals" as one Islamic historian called them; "beasts unrestrained save by drunkenness or terror," as another put it—with the typical prejudice of urban Arabs), rose up in arms and founded a kingdom of his own in the North. This man, named Ghailan, looked like a potential savior to the Andalusians of Rabat. They staged an uprising, and beseiged "Prince" Abdullah in the Casbah. The Dala'iyya master M. al-Hajj sent an army to relieve his son, but the army was defeated by Ghailan in June 1660. Abdullah however held on gamely in the Casbah for another year, helped by a shipment of supplies sent by the English governor of Tangiers. At last, in June 1661, he ran out of food and had to surrender the castle.

By this time the Andalusians had come to distrust Ghailan as much as they'd disliked the Dala'iyya—more, in truth. Despite the fact that they'd just run the Dala'iyya out of town, they decided to profess renewed loyalty to the regime in order to stave off Ghailan, lest he prove a worse master. For four years they played hard-to-get, but finally in 1664 capitulated to Ghailan and agreed to pay him the dreaded 10 percent.

Finally, in 1668, the last vestiges of Salé's freedom were wiped out by the rise of the Alawite Dynasty under its Sultan Moulay Raschid, who succeeded in reuniting the whole country for the first time since 1603. The Alawite Sultan had no intention of putting an end to the highly profitable holy war of the Bon Regreg against Europe, and promised the corsairs his protection. Thus, although the Republic had vanished, piracy survived—for awhile. Unfortunately the Alawites had huge appetites, and little by little increased the "bite" from 10% to well over half. Eventually the corsairs realized that decent profits were no longer possible. The Moorish pirates stayed on to become captains in the Sultan's "Navy," and perhaps some of the Renegadoes did the same. Others, perhaps, were tempted to move on, to the Caribbean, or to Madagascar, where the pirate scene now began to flourish. The later history of Salé, nor the later history of Barbary in general. With the passing of the Republic we lose sight of our Renegadoes.[32]

Salé, Algiers, Triopoli and Tunis were all bases for the corsairs, a bunch of pirates who, like the lost Templar fleet, were not on particularly good terms with the Vatican and her allied nations along the north coast of the Mediterranean. In fact, the corsairs didn't like ships from any Christian nation!

§§§

Thomas Jefferson, Stephen Decatur, and the Corsairs
In 1801 the newly independent United States, whose ships had also come under attack, launched the Tripolitan War (1801-05) against the pirate utopias of the corsairs. Essentially, the Tripolitan War was an effort by the United States to end extortion payments to Barbary Coast pirates. The newly created Templar-Mason republic was to send its fleet of privateers to North Africa to battle other pirates. Indeed, in a scenario similar to that of Captain Kidd's, where he was a privateer commissioned to hunt pirates, so was the U.S. navy a group of buccaneers with some dried beef and cutlasses ready to take on the Barbary Coast corsairs.

Prior to 1801 the United States, along with the European powers, routinely negotiated extortionate treaties and paid tribute to the North African Barbary States (Tripoli, Algiers, Morocco, and Tunis) in exchange for the safe passage of merchant vessels through the Mediterranean. In that year, the pasha of Tripoli repudiated his treaty with the United States and declared war.

President Thomas Jefferson dispatched the navy's Mediterranean squadron, and the U.S. show of force discouraged the other Barbary States from backing the pasha of Tripoli and his demands for ransoms and tribute. Jefferson's squadron comprised three Masonic naval commanders: Commodore Edward Preble, William Eaton and Stephen Decatur.

Commissioned a midshipman in 1798, Stephen Decatur served in the West Indies during the Quasi-War with France during the period 1798-1800. In the Tripolitan War, Decatur led a small band of sailors into Tripoli harbor on Feb. 16, 1804. There they boarded and set fire to the captured American frigate *Philadelphia* before escaping with only one man wounded. The British admiral Horatio Nelson hailed the exploit as the "most bold and daring act of the age," and Decatur was promoted to captain.

Decatur further distinguished himself in the War of 1812. In the frigate *United States*, he defeated the British frigate *Macedonian* in one of the war's most renowned single-ship engagements on October 25, 1812. He was subsequently unable to leave American waters because of the British blockade. When he tried to run the blockade out of New York in January 1815, his ship was captured by a British squadron.

By this time the war had ended, however, and he was repatriated. In June 1815 he returned to the Mediterranean and dictated terms of peace to Algiers, Tunis, and Tripoli, thereby concluding the last American war with the Barbary States. At a banquet celebrating this achievement, Decatur chauvinistically toasted: "Our country! In her intercourse with foreign nations may she always be in the right; but our country, right or wrong." A practiced and ardent duelist, Decatur accepted a challenge from James Barron, a suspended captain on whose court martial Decatur had sat, and was killed in the duel.

Commodore Edward Preble was from Falmouth (now Portland), Maine, born on Aug. 15, 1761. He served in the Massachusetts navy during the American Revolution and in the U.S. Navy during the Quasi-War (1798-1800) with France. As captain of the frigate *Essex* in 1800, he escorted a convoy of American merchantmen to the East Indies; this cruise marked the first appearance of a U.S. warship in the waters east of the Cape of Good Hope. Appointed Mediterranean commander in 1803, Preble energetically prosecuted the war against Tripoli. As commander, he ordered the blockade of Tripoli, and endorsed Stephen Decatur's bold plan to enter Tripoli harbor and burn the captured U.S. frigate *Philadelphia*.

Our third officer on the scene, William Eaton, launched an overland campaign after the burning of the *Philadelphia*. The combination of the successful rescue of the captives, the burning of the prize ship, and, of all things, an overland American army wreaking havoc around the Magreb of North Africa caused the local ruler of Tripoli to desire a swift end to the disruptive activities of the Barbary corsairs.

In 1805 the pasha signed a treaty requiring ransom for American prisoners plus occasional presents, but ending annual tribute. The Americans paid the ransom and took the freed Americans back to New England.

A final resolution of the problems with the Barbary States came in 1815 when Congress, responding to piratical attacks on U.S. commerce, declared war on Algiers. Decatur rushed a squadron to the Mediterranean, where he dictated peace on American terms to Algiers and then to Tunis and Tripoli. Shortly after Decatur's departure from Algiers, it was bombarded by an Anglo-Dutch fleet in early 1816.

The American payment of tribute and the intermittent wars with the Barbary States ended, and the United States established a permanent naval squadron in the Mediterranean. Today, by a certain irony, the American naval contingent in the Mediterranean is the largest navy in the region.

And so we come full circle from the ancient seafarers, to the early days of piracy with the Phoenicians, to the lost fleet of the Knights Templar, to the Masonic pirates who created the United States and her magnificent navy. To all those scurvy dogs and their motley crews who sail the Seven Seas, I say, "It's a pirate's life for me!"

American ships prepare to leave port in 1776.

A British cartoon in *Punch* about American pirates and the Jolly Roger.

General Lafayette.

Left: Stephen Decatur's Masonic emblem. Right: The *Bonhomme Richard*.

John Paul Jones, pirate founder of the American Navy.

John Paul Jones battles the *Serapis* with cannon broadsides.

John Paul Jones, pirate founder of the American Navy.

John Paul Jones battles the *Serapis* with cannon broadsides.

The Barbarossa brothers.

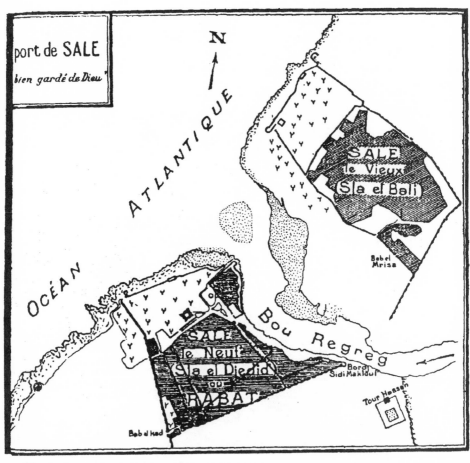

The independent pirate utopia, Salé, in Morocco.

George Washington Depicted as a Freemason.

Bibliography and Footnotes

1. **Holy Blood, Holy Grail**, Michael Baigent, Richard Leigh & Henry Lincoln, 1982, Jonathan Cape, London (published in the U.K. as The Holy Blood and the Holy Grail).
2. **The Messianic Legacy**, Michael Baigent, Richard Leigh & Henry Lincoln, 1985, Jonathan Cape, London.
3. **The Temple and the Lodge,** Michael Baigent & Richard Leigh, 1989, Jonathan Cape, London.
4. **Irish Druids and Old Irish Religions**, James Bonwick, 1894. Dorset reprint 1986.
5. **Irish Symbols of 3500 B.C.**, N.L. Thomas, 1988, Mericer Press, Dublin.
6. **The History of the Knights Templars**, Charles G. Addison, 1842, London (reprinted by Adventures Unlimited Press, Kempton, IL.
7. **A History of Secret Societies**, Arkon Daraul, 1962, Citadel Press, NY.
8. **The Mysteries of Chartres Cathedral**, Louis Charpentier, 1975, Avon Books, New York. 1966, Robert Lafont, Paris.
9. **Holy Grail Across the Atlantic**, Michael Bradley, 1988, Hounslow Press, Willowdale, Ontario.
10. **The Columbus Conspiracy**, Michael Bradley, 1992,, A & B Publishers, Brooklyn.
11. **Prince Henry Sinclair**, Frederick Pohl, 1974, Clarkson Potter Publisher, New York.
12. **The Sword and the Grail**, Andrew Sinclair, 1992, Crown, New York.
13. **The Templar Legacy & the Masonic Inheritance Within Rosslyn Chapel**, Tim Wallace-Murphy, 1993, Friends of Rosslyn, Rosslyn, Scotland.
14. **Oak Island Gold**, William S. Crooker, 1993, Nimbus Publishing, Halifax, Nova Scotia.
15. **Tracking Treasure**, William S. Crooker, 1998, Nimbus Publishing, Halifax, Nova Scotia.

16. **America BC**, Barry Fell, 1976, Demeter Press, NYC.
17. **The Pirate Hunter,** Richard Zacks, 2002, Hyperion Books, New York.
18. **The Templars, Knights of God**, Edward Burman,1986, Destiny Books, Rochester, Vermont.
19. **Lost Cities of North & Central America**, David Hatcher Childress, 1992, Adventures Unlimited Press, Kempton, IL.
20. **The Prize of the All the Oceans**, Glyn Williams, 1999, Viking, New York.
21. **The Search For the Stone of Destiny**, Pat Gerber, 1992, Canongate Press, Edinburgh.
22. **The Knights Templar and Their Myth**, Peter Partner, 1987, Destiny Books, Rochester, Vermont.
23. **The Knights Templar**, Stephen Howarth, 1982, MacMillan Publishing, New York.
24. **Under the Black Flag**, David Cordingly, 1995, Harcourt Brace & Co., New York.
25. **Bloodline of the Holy Grail**, Laurence Gardner, 1996, Element Books, London.
26. **Genesis of the Grail Kings**, Laurence Gardner, 1998, Element Books, London.
27. **Realm of the Ring Lords**, Laurence Gardner, 2000, Media Quest, London.
28. **King Solomon's Temple**, E. Raymond Capt, 1979, Artisan Sales, Thousand Oaks, CA.
29. **A General History of Pirates**, Captain Charles Johnson, 1998, Lyons Press, NYC (reissue).
30. **Secrets of the Lost Races**, Rene Noorbergen, 1977, Barnes & Noble Publishers, New York.
31. **Plato Prehistorian**, Mary Settegast, 1990, Lindisfarne Press, Hudson, NY.
32. **Pirate Utopias: Moorish Corairs & European Renegadoes**, Peter Lamborn Wilson, 1995, Autonomedia, Brooklyn, NY.
33. **The Templars' Legacy in Montreal, the New Jerusalem**, Francine Bernier, 2001, Frontier Publishing-Adventures Unlimited, Enkhuizen & Kempton, Holland & Illinois.
34. **The Templars and the Assassins**, James Wasserman, 2001, Inner Traditions, Rochester, VT.
35. **Homer's Secret Illiad**, Florence and Kenneth Wood, 1999, John Murray Co., London.
36. **Maps of the Ancient Sea Kings**, Charles Hapgood, 1970, Adventures Unlimited Press, Kempton, Illinois.
37. **Sources of the Grail**, John Matthews, 1996, Floris Publishers, Edinburgh.
38. **The Templar Treasure At Gisors**, Jean Markale, 1986 (2003 English

edition), Inner Traditions, Rochester, VT.
39. **Early Man and the Ocean**, Thor Heyerdahl, 1978, Doubleday & Co., Garden City, New Jersey.
40. **Sailing To Paradise, The Discovery of the Americas by 7000 B.C.**, Jim Bailey, 1994, Simon & Schuster, New York.
41. **The Holy Land of Scotland**, Barry Dunford, 1996. Sacred Connection, Perthshire, Scotland.
42. **Dungeon, Fire and Sword**, John A. Robinson, 1991, Brockhampton Press, London.
43. **Pirates, Privateers, and Rebel Raiders**, 2000, Lindley S. Butler, U. of N. Carolina Press, Chapel Hill.
44. **The Lost Treasure of the Knights Templar**, Steven Sora, 1999, Inner Traditions, Rochester, VT.
45. **Kingdom of the Ark**, Lorraine Evans, 2000, Simon & Schuster, London.
46. **The Merovingian Kingdoms 450-751**, Ian Wood, 1994, Longman, Essex.
47. **The Templar Revelation**, Lynn Picknett and Clive Prince, 1997, Bantam, London.
48. **The Knights of the Order**, Ernle Bradford, 1972, Barnes & Noble, New York.
49. **6000 Years of Seafaring**, Orville Hope, 1983, Hope Press, Gastonia, NC.
50. **America's Ancient Treasures**, Franklin & Mary Folsom, 1971, University of New Mexico Press, Albuquerque.
51. **Conquest By Man,** Paul Herrmann, 1974, Souvenir Press, London.
52. **They All Discovered America**, Boland, 1961, Doubleday, NYC.
53. **Men Who Dared the Sea**, Gardner Soule, 1976, Thomas Crowell Co., NYC.
54. **Technology In the Ancient World**, Henry Hodges, 1970, Marlboro Books, London.
55. **Rosslyn: Guardians of the Secrets of the Holy Grail**, Tim Wallace-Murphy, 1999, Element Books, London.
56. **The Buccaneers of America,** Alexander O. Exquemelin, 1969, Penguin Books, London.
57. **Under the Black Flag,** Don C. Seitz, 1925, Dial Press, New York.
58. **Pirates and Buccaneers**, Peter F. Copeland, 1977, Dover Publications, New York.
59. **1421: The Year China Discovered America**, Gavin Menzies, 2002, Random House, New York.
56. **Saga America**, Barry Fell, 1980, New York Times Books, New York.
57. **Lost Cities & Ancient Mysteries of South America**, David Hatcher Childress, 1987, AUP, Kempton, Illinois.
58. **Lost Cities & Ancient Mysteries of Africa & Arabia**, David Hatcher

Childress, 1990, AUP, Kempton, Illinois.

59. **Lost Cities of North & Central America**, David Hatcher Childress, 1994, AUP, Kempton, Illinois.

60. **Atlantic Crossings Before Columbus**, Frederick Pohl, 1961, Norton, New York.

61. **Pale Ink**, Henriette Mertz, 1953 (1972, revised, 2nd edition), Swallow Press, Chicago.

62. **Atlantis, Dwelling Place of the Gods**, Henriette Mertz, 1976, Swallow Press, Chicago.

63. **The Wine Dark Sea**, Henriette Mertz, 1964, Swallow Press, Chicago.

64. **The Mystic Symbol**, Henriette Mertz, 1986, Global Books, Chicago.

65. **Bronze Age America**, Barry Fell, 1982, Little, Brown & Co., Boston

66. **Viking Explorers & the Columbus Fraud**, Wilford Raymond Anderson, 1981, Valhalla Press, Chicago.

67. **Admiral of the Ocean Sea**, Samuel Eliot Morison, 1942, Little, Brown, NYC.

68. **Sails of Hope**, Simon Wiesenthal, 1973, Macmillan, NYC.

69. **Encyclopedia Brittanica**, 1973 editi on, Volume 6.

70. **Universal Jewish Encyclopedia**, 1969 edition, Volume 3.

71. **Vanished Cities**, Hermann & Georg Schreiber, 1957, Alfred Knopf, New York.

72. **Legendary Islands of the Atlantic,** William Babcock, 1922, New York.

73. **Hidden Worlds,** Van der Meer & Moerman, 1974, Souvenir Press, London.

74. **Mysterious America**, Loren Coleman, 1983, Faber & Faber, London & Boston.

75. **Calalus**, Cyclone Covey, 1975, Vantage Press, New York.

76. **No Longer On the Map,** Raymond H. Ramsay, 1972, Viking, New York.

77. **World of the Odd & Awesome**, Charles Berlitz, 1991, Fawcett, New York.

78. **Masonic Membership of the Founding Fathers,** Brother Ronald E. Heaton, 1965, Masonic Service Association, Silver Spring, Maryland.

79. **The Stone Puzzle of Rosslyn Chapel**, Philip Coppens, 2002, Frontier Publishing-Adventures Unlimited, Enkhuizen & Kempton, Holland and Illinois.

80. **Piracy In the Ancient World**, Henry A. Ormerod, 1924, University of Liverpool Press.

81. **Buccaneers and Pirates of Our Coasts**, Frank R. Stockton, 1896, Looking Glass Library, New York.

82. **When China Ruled the Seas**, Louise Levathes, 1994, Simon & Schuster, NYC.

83. **Men Out of Asia**, Harold Gladwin, 1947, McGraw-Hill, NYC.

84. **Engineering In the Ancient World**, J.G. Landels, 1978, U. of California

Press, Berkeley.

85. **Archaeology Beneath the Sea**, George F. Bass, 1975, Walker & Co., NYC.

86. **Cities In the Sea**, Nichols C. Fleming, 1971, Doubleday & Co., Garden City, NJ.

87. **4,000 Years Under the Sea**, Philippe Diolé, 1952, Julian Messner, Inc, NY.

88. **Did the Phoenicians Discover America?**, Thomas Johnston, 1913, James Nesbit & Co. London.

89. **What the Bible Really Says**, Manfred Barthel, 1982, Souvenir Press, London.

90. **Secrets of the Bible Seas**, Alexander Flinder, 1985, Severn House Publishers, London.

91. **Lost Cities of Ancient Lemuria & the Pacific**, David Hatcher Childress, 1988, Adventures Unlimited Press, Kempton, Illinois.

92. **Riddles in History**, Cyrus H. Gordon, 1974, Crown Publishers, NYC.

93. **Lost Worlds**, Alistair Service, 1981, Arco Publishing, New York.

94. **The World's Last Mysteries**, Readers Digest, 1976, Pleasantville, NY.

95. **Grolier Multimedia Encyclopedia**, 1997, Danbury, CT.

96. **European Treaties Bearing on the History of the United States and its Dependencies to 1648**, Frances Gardiner Davenport, ed., 1917, online.

CONSPIRACY & HISTORY

TEMPLARS' LEGACY IN MONTREAL
The New Jerusalem
by Francine Bernier

Designed in the 17th century as the New Jerusalem of the Christian world, the people behind the scene in turning this dream into reality were the Société de Notre-Dame, half of whose members were in the elusive Compagnie du Saint-Sacrement. They took no formal vows and formed the interior elitist and invisible "heart of the church" following a "Johannite" doctrine of the Essene tradition, where men and women were considered equal apostles. The book reveals the links between Montreal and: John the Baptist as patron saint; Melchizedek, the first king-priest and a father figure to the Templars and the Essenes; Stella Maris, the Star of the Sea from Mount Carmel; the Phrygian goddess Cybele as the androgynous Mother of the Church; St. Blaise, the Armenian healer or "Therapeut"- the patron saint of the stonemasons and a major figure to the Benedictine Order and the Templars; the presence of two Black Virgins; an intriguing family coat of arms with twelve blue apples; and more.
352 PAGES. 6X9 PAPERBACK. ILLUSTRATED. BIBLIOGRAPHY. $21.95. CODE: TLIM

THE STONE PUZZLE OF ROSSLYN CHAPEL
by Philip Coppens

Rosslyn Chapel is revered by Freemasons as a vital part of their history, believed by some to hold evidence of pre-Columbian voyages to America, assumed by others to hold important relics, from the Holy Grail to the Head of Christ, the Scottish chapel is a place full of mystery. The history of the chapel, its relationship to freemasonry and the family behind the scenes, the Sinclairs, is brought to life, incorporating new, previously forgotten and heretofore unknown evidence. Significantly, the story is placed in the equally enigmatic landscape surrounding the chapel, which includes features from Templar commanderies to prehistoric markings, from an ancient kingly site to the South to Arthur's Seat directly north of the chapel. The true significance and meaning of the chapel is finally unveiled: it is a medieval stone book of esoteric knowledge "written" by the Sinclair family, one of the most powerful and wealthy families in Scotland, chosen patrons of Freemasonry.
124 PAGES. 6X9 PAPERBACK. ILLUSTRATED. $12.00. CODE: SPRC

NOSTRADAMUS AND THE LOST TEMPLAR LEGACY
by Rudy Cambier

Rudy Cambier's decade-long research and analysis of the verses of Nostradamus' "prophecies" has shown that the language of those verses does not belong in the 16th century, nor in Nostradamus' region of Provence. The language spoken in the verses belongs to the medieval times of the 14th Century, and the Belgian borders. The documents known as Nostradamus' prophecies were not written ca. 1550 by the French "visionary" Michel de Nostradame. Instead, they were composed between 1323 and 1328 by a Cistercian monk, Yves de Lessines, prior of the abbey of Cambron, on the border between France and Belgium. According to the author, these documents reveal the location of a Templar treasure. This key allowed Cambier to translate the "prophecies." But rather than being confronted with a series of cataclysms and revelations of future events, Cambier discovered a possibly even more stunning secret. Yves de Lessines had waited for many years for someone called "l'attendu," the expected one. This person was supposed to come to collect the safeguarded treasures of the Knights Templar, an organization suppressed in 1307. But no-one came. Hence, the prior decided to impart the whereabouts and nature of the treasure in a most cryptic manner in verses.
204 PAGES. 6X9 PAPERBACK. ILLUSTRATED. BIBLIOGRAPHY. $17.95. CODE: NLTL

THE DIMENSIONS OF PARADISE
The Proportions & Symbolic Numbers of Ancient Cosmology
by John Michell

The Dimensions of Paradise were known to ancient civilizations as the harmonious numerical standards that underlie the created world. John Michell's quest for these standards provides vital clues for understanding: The dimensions and symbolism of Stonehenge; The plan of Atlantis and reason for its fall; The numbers behind the sacred names of Christianity; The form of St. John's vision of the New Jerusalem; The name of the man with the number 666; The foundation plan of Glastonbury and other sanctuaries and how these symbols suggest a potential for personal, cultural and political regeneration in the 21st century.
220 PAGES. 6X9 PAPERBACK. ILLUSTRATED. BIBLIOGRAPHY. INDEX. $16.95. CODE: DIMP

THE HISTORY OF THE KNIGHTS TEMPLARS
by Charles G. Addison, introduction by David Hatcher Childress

Chapters on the origin of the Templars, their popularity in Europe and their rivalry with the Knights of St. John, later to be known as the Knights of Malta. Detailed information on the activities of the Templars in the Holy Land, and the 1312 AD suppression of the Templars in France and other countries, which culminated in the execution of Jacques de Molay and the continuation of the Knights Templars in England and Scotland; the formation of the society of Knights Templars in London; and the rebuilding of the Temple in 1816. Plus a lengthy intro about the lost Templar fleet and its connections to the ancient North American sea routes.
395 PAGES. 6X9 PAPERBACK. ILLUSTRATED. $16.95. CODE: HKT

SAUNIER'S MODEL AND THE SECRET OF RENNES-LE-CHATEAU
The Priest's Final Legacy
by André Douzet

Berenger Saunière, the enigmatic priest of the French village of Rennes-le-Château, is rumored to have found the legendary treasure of the Cathars. But what became of it? In 1916, Saunier created his ultimate clue: he went to great expense to create a model of a region said to be the Calvary Mount, indicating the "Tomb of Jesus." But the region on the model does not resemble the region of Jerusalem. Did Saunière leave a clue as to the true location of his treasure? And what is that treasure? After years of research, André Douzet discovered this model— the only real clue Saunière left behind as to the nature and location of his treasure—and the possible tomb of Jesus.
116 PAGES. 6X9 PAPERBACK. ILLUSTRATED. BIBLIOGRAPHY. $12.00. CODE: SMOD

ARKTOS
The Myth of the Pole in Science, Symbolism, and Nazi Survival
by Joscelyn Godwin

A scholarly treatment of catastrophes, ancient myths and the Nazi Occult beliefs. Explored are the many tales of an ancient race said to have lived in the Arctic regions, such as Thule and Hyperborea. Progressing onward, the book looks at modern polar legends including the survival of Hitler, German bases in Antarctica, UFOs, the hollow earth, Agartha and Shambala, more.
220 PAGES. 6X9 PAPERBACK. ILLUSTRATED. $16.95. CODE: ARK

THE GIZA DEATH STAR
The Paleophysics of the Great Pyramid & the Military Complex at Giza
by Joseph P. Farrell

Physicist Joseph Farrell's amazing book on the secrets of Great Pyramid of Giza. *The Giza Death Star* starts where British engineer Christopher Dunn leaves off in his 1998 book, *The Giza Power Plant*. Was the Giza complex part of a military installation over 10,000 years ago? Chapters include: An Archaeology of Mass Destruction, Thoth and Theories; The Machine Hypothesis; Pythagoras, Plato, Planck, and the Pyramid; The Weapon Hypothesis; Encoded Harmonics of the Planck Units in the Great Pyramid; High Freгquency Direct Current "Impulse" Technology; The Grand Gallery and its Crystals: Gravito-acoustic Resonators; The Other Two Large Pyramids; the "Causeways," and the "Temples"; A Phase Conjugate Howitzer; Evidence of the Use of Weapons of Mass Destruction in Ancient Times; more.
290 PAGES. 6x9 PAPERBACK. ILLUSTRATED. $16.95. CODE: GDS

THE GIZA DEATH STAR DEPLOYED
The Physics & Engineering of the Great Pyramid
by Joseph P. Farrell

Physicist Joseph Farrell's amazing sequel to *The Giza Death Star* which takes us from the Great Pyramid to the asteroid belt and the so-called Pyramids of Mars. Farrell expands on his thesis that the Great Pyramid was a chemical maser, designed as a weapon and eventually deployed—with disastrous results to the solar system. Includes: Exploding Planets: The Movie, the Mirror, and the Model; Dating the Catastrophe and the Compound; A Brief History of the Exoteric and Esoteric Investigations of the Great Pyramid; No Machines, Please!; The Stargate Conspiracy; The Scalar Weapons; Message or Machine?; A Tesla Analysis of the Putative Physics and Engineering of the Giza Death Star; Cohering the Zero Point, Vacuum Energy, Flux: Synopsis of Scalar Physics and Paleophysics; Configuring the Scalar Pulse Wave; Inferred Applications in the Great Pyramid; Quantum Numerology, Feedback Loops and Tetrahedral Physics; and more.
290 PAGES. 6x9 PAPERBACK. ILLUSTRATED. BIBLIOGRAPHY. INDEX. $16.95. CODE: GDSD

PIRATES & THE LOST TEMPLAR FLEET
The Secret Naval War Between the Templars & the Vatican
by David Hatcher Childress

The lost Templar fleet was originally based at La Rochelle in southern France, but fled to the deep fiords of Scotland upon the dissolution of the Order by King Phillip. This banned fleet of ships was later commanded by the St. Clair family of Rosslyn Chapel (birthplace of Free Masonry). St. Clair and his Templars made a voyage to Canada in the year 1398 AD, nearly 100 years before Columbus! Chapters include: 10,000 Years of Seafaring; The Knights Templar & the Crusades; The Templars and the Assassins; The Lost Templar Fleet and the Jolly Roger; Maps of the Ancient Sea Kings; Pirates, Templars and the New World; Christopher Columbus—Secret Templar Pirate?; Later Day Pirates and the War with the Vatican; Pirate Utopias and the New Jerusalem; more.
320 PAGES. 6x9 PAPERBACK. ILLUSTRATED. BIBLIOGRAPHY. $16.95. CODE: PLTF

CLOAK OF THE ILLUMINATI
Secrets, Transformations, Crossing the Star Gate
by William Henry

Thousands of years ago the stargate technology of the gods was lost. Mayan Prophecy says it will return by 2012, along with our alignment with the center of our galaxy. In this book: Find examples of stargates and wormholes in the ancient world; Examine myths and scripture with hidden references to a stargate cloak worn by the Illuminati, including Mari, Nimrod, Elijah, and Jesus; See rare images of gods and goddesses wearing the Cloak of the illuminati; Learn about Saddam Hussein and the secret missing library of Jesus; Uncover the secret Roman-era eugenics experiments at the Temple of Hathor in Denderah, Egypt; Explore the duplicate of the Stargate Pillar of the Gods in the Illuminists' secret garden in Nashville, TN; Discover the secrets of manna, the food of the angels; Share the lost Peace Prayer posture of Osiris, Jesus and the Illuminati; more. Chapters include: Seven Stars Under Three Stars; The Long Walk; Squaring the Circle; The Mill of the Host; The Miracle Garment; The Fig; Nimrod: The Mighty Man; Nebuchadnezzar's Gate; The New Mighty Man; more.
238 PAGES. 6x9 PAPERBACK. ILLUSTRATED. BIBLIOGRAPHY. INDEX. $16.95. CODE: COIL

THE CHRONOLOGY OF GENESIS
A Complete History of Nefilim
by Neil Zimmerer

Follow the Nefilim through the Ages! This is a complete history of Genesis, the gods and the history of Earth — before the gods were destroyed by their own creations more than 2500 years ago! Zimmerer presents the most complete history of the Nefilim ever developed — from the Sumerian Nefilim kings through the Nefilim today. He provides evidence of extraterrestrial Nefilim monuments, and includes fascinating information on pre-Nefilim man-apes and man-apes of the world in the present age. Includes the following subjects and chapters: Creation of the Universe; Evolution: The Greatest Mystery; Who Were the Nefilim?; Pre-Nefilim Man-Apes; Man-Apes of the World—Present Age; Extraterrestrial Nefilim Monuments; The Nefilim Today; All the Sumerian Nefilim Kings listed in chronological order, more. A book not to be missed by researchers into the mysterious origins of mankind.
244 PAGES. 6x9 PAPERBACK. ILLUSTRATED. REFERENCES. $16.95. CODE: CGEN

LEY LINE & EARTH ENERGIES
An Extraordinary Journey into the Earth's Natural Energy System
by David Cowan & Anne Silk

The mysterious standing stones, burial grounds and stone circles that lace Europe, the British Isles and other areas have intrigued scientists, writers, artists and travellers through the centuries. They pose so many questions: Why do some places feel special? How do ley lines work? How did our ancestors use Earth energy to map their sacred sites and burial grounds? How do ghosts and poltergeists interact with Earth energy? How can Earth spirals and black spots affect our health? This exploration shows how natural forces affect our behavior, how they can be used to enhance our health and well being, and ultimately, how they bring us closer to penetrating one of the deepest mysteries being explored. A fascinating and visual book about subtle Earth energies and how they affect us and the world around them.
368 PAGES. 6x9 PAPERBACK. ILLUSTRATED. BIBLIOGRAPHY. INDEX. $18.95. CODE: LLEE

ATLANTIS STUDIES

MAPS OF THE ANCIENT SEA KINGS
Evidence of Advanced Civilization in the Ice Age
by Charles H. Hapgood
Charles Hapgood's classic 1966 book on ancient maps produces concrete evidence of an advanced world-wide civilization existing many thousands of years before ancient Egypt. He has found the evidence in the Piri Reis Map that shows Antarctica, the Hadji Ahmed map, the Oronteus Finaeus and other amazing maps. Hapgood concluded that these maps were made from more ancient maps from the various ancient archives around the world, now lost. Not only were these unknown people more advanced in mapmaking than any people prior to the 18th century, it appears they mapped all the continents. The Americas were mapped thousands of years before Columbus. Antarctica was mapped when its coasts were free of ice.
316 PAGES. 7x10 PAPERBACK. ILLUSTRATED. BIBLIOGRAPHY & INDEX. $19.95. CODE: MASK

PATH OF THE POLE
Cataclysmic Pole Shift Geology
by Charles Hapgood
Maps of the Ancient Sea Kings author Hapgood's classic book *Path of the Pole* is back in print! Hapgood researched Antarctica, ancient maps and the geological record to conclude that the Earth's crust has slipped in the inner core many times in the past, changing the position of the pole. *Path of the Pole* discusses the various "pole shifts" in Earth's past, giving evidence for each one, and moves on to possible future pole shifts. Packed with illustrations, this is the sourcebook for many other books on cataclysms and pole shifts.
356 PAGES. 6x9 PAPERBACK. ILLUSTRATED. $16.95. CODE: POP.

ATLANTIS & THE POWER SYSTEM OF THE GODS
Mercury Vortex Generators & the Power System of Atlantis
by David Hatcher Childress and Bill Clendenon
Atlantis and the Power System of the Gods starts with a reprinting of the rare 1990 book *Mercury: UFO Messenger of the Gods* by Bill Clendenon. Clendenon takes on an unusual voyage into the world of ancient flying vehicles, strange personal UFO sightings, a meeting with a "Man In Black" and then to a centuries-old library in India where he got his ideas for the diagrams of mercury vortex engines. The second part of the book is Childress' fascinating analysis of Nikola Tesla's broadcast system in light of Edgar Cayce's "Terrible Crystal" and the obelisks of ancient Egypt and Ethiopia. Includes: Atlantis and its crystal power towers that broadcast energy; how these incredible power stations may still exist today; inventor Nikola Tesla's nearly identical system of power transmission; Mercury Proton Gyros and mercury vortex propulsion; more. Richly illustrated, and packed with evidence that Atlantis not only existed—it had a world-wide energy system more sophisticated than ours today.
246 PAGES. 6x9 PAPERBACK. ILLUSTRATED. $15.95. CODE: APSG

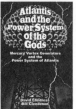

ATLANTIS IN AMERICA
Navigators of the Ancient World
by Ivar Zapp and George Erikson
This book is an intensive examination of the archeological sites of the Americas, an examination that reveals civilization has existed here for tens of thousands of years. Zapp is an expert on the enigmatic giant stone spheres of Costa Rica, and maintains that they were sighting stones similar to those found throughout the Pacific as well as in Egypt and the Middle East. They were used to teach star-paths and sea navigation to the world-wide navigators of the ancient world. While the Mediterranean and European regions "forgot" world-wide navigation and fought wars, the Mesoamericans of diverse races were building vast interconnected cities without walls. This Golden Age of ancient America was merely a myth of suppressed history—until now. Profusely illustrated, chapters are on Navigators of the Ancient World; Pyramids & Megaliths: Older Than You Think; Ancient Ports and Colonies; Cataclysms of the Past; Atlantis: From Myth to Reality; The Serpent and the Cross: The Loss of the City States; Calendars and Star Temples; and more.
360 PAGES. 6x9 PAPERBACK. ILLUSTRATED. BIBLIOGRAPHY & INDEX. $17.95. CODE: AIA

FAR-OUT ADVENTURES   *REVISED EDITION*
The Best of World Explorer Magazine
This is a compilation of the first nine issues of *World Explorer* in a large-format paperback. Authors include: David Hatcher Childress, Joseph Jochmans, John Major Jenkins, Deanna Emerson, Katherine Routledge, Alexander Horvat, Greg Deyermenjian, Dr. Marc Miller, and others. Articles in this book include Smithsonian Gate, Dinosaur Hunting in the Congo, Secret Writings of the Incas, On the Trail of the Yeti, Secrets of the Sphinx, Living Pterodactyls, Quest for Atlantis, What Happened to the Great Library of Alexandria?, In Search of Seamonsters, Egyptians in the Pacific, Lost Megaliths of Guatemala, the Mystery of Easter Island, Comacalco: Mayan City of Mystery, Professor Wexler and plenty more.
580 PAGES. 8x11 PAPERBACK. ILLUSTRATED. REVISED EDITION. $25.00. CODE: FOA

RETURN OF THE SERPENTS OF WISDOM
by Mark Amaru Pinkham
According to ancient records, the patriarchs and founders of the early civilizations in Egypt, India, China, Peru, Mesopotamia, Britain, and the Americas were the Serpents of Wisdom—spiritual masters associated with the serpent—who arrived in these lands after abandoning their beloved homelands and crossing great seas. While bearing names denoting snake or dragon (such as Naga, Lung, Djedhi, Amaru, Quetzalcoatl, Adder, etc.), these Serpents of Wisdom oversaw the construction of magnificent civilizations within which they and their descendants served as the priest kings and as the enlightened heads of mystery school traditions. *The Return of the Serpents of Wisdom* recounts the history of these "Serpents"—where they came from, why they came, the secret wisdom they disseminated, and why they are returning now.
400 PAGES. 6x9 PAPERBACK. ILLUSTRATED. REFERENCES. $16.95. CODE: RSW

24 hour credit card orders—call: 815-253-6390 fax: 815-253-6300
email: auphq@frontiernet.net   www.adventuresunlimitedpress.com   www.wexclub.com

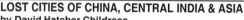

LOST CITIES

LOST CITIES OF ATLANTIS, ANCIENT EUROPE & THE MEDITERRANEAN
by David Hatcher Childress

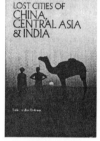

Atlantis! The legendary lost continent comes under the close scrutiny of maverick archaeologist David Hatcher Childress in this sixth book in the internationally popular *Lost Cities* series. Childress takes the reader in search of sunken cities in the Mediterranean; across the Atlas Mountains in search of Atlantean ruins; to remote islands in search of megalithic ruins; to meet living legends and secret societies. From Ireland to Turkey, Morocco to Eastern Europe, and around the remote islands of the Mediterranean and Atlantic, Childress takes the reader on an astonishing quest for mankind's past. Ancient technology, cataclysms, megalithic construction, lost civilizations and devastating wars of the past are all explored in this book. Childress challenges the skeptics and proves that great civilizations not only existed in the past, but the modern world and its problems are reflections of the ancient world of Atlantis.
524 PAGES. 6x9 PAPERBACK. ILLUSTRATED. BIBLIOGRAPHY & INDEX. $16.95. CODE: MED

LOST CITIES OF CHINA, CENTRAL INDIA & ASIA
by David Hatcher Childress
Like a real life "Indiana Jones," maverick archaeologist David Childress takes the reader on an incredible adventure across some of the world's oldest and most remote countries in search of lost cities and ancient mysteries. Discover ancient cities in the Gobi Desert; hear fantastic tales of lost continents, vanished civilizations and secret societies bent on ruling the world; visit forgotten monasteries in forbidding snow-capped mountains with strange tunnels to mysterious subterranean cities! A unique combination of far-out exploration and practical travel advice, it will astound and delight the experienced traveler or the armchair voyager.
429 PAGES. 6x9 PAPERBACK. ILLUSTRATED. FOOTNOTES & BIBLIOGRAPHY. $14.95. CODE: CHI

LOST CITIES OF ANCIENT LEMURIA & THE PACIFIC
by David Hatcher Childress
Was there once a continent in the Pacific? Called Lemuria or Pacifica by geologists, Mu or Pan by the mystics, there is now ample mythological, geological and archaeological evidence to "prove" that an advanced and ancient civilization once lived in the central Pacific. Maverick archaeologist and explorer David Hatcher Childress combs the Indian Ocean, Australia and the Pacific in search of the surprising truth about mankind's past. Contains photos of the underwater city on Pohnpei; explanations on how the statues were levitated around Easter Island in a clockwise vortex movement; tales of disappearing islands; Egyptians in Australia; and more.
379 PAGES. 6x9 PAPERBACK. ILLUSTRATED. FOOTNOTES & BIBLIOGRAPHY. $14.95. CODE: LEM

LOST CITIES OF NORTH & CENTRAL AMERICA
by David Hatcher Childress
Down the back roads from coast to coast, maverick archaeologist and adventurer David Hatcher Childress goes deep into unknown America. With this incredible book, you will search for lost Mayan cities and books of gold, discover an ancient canal system in Arizona, climb gigantic pyramids in the Midwest, explore megalithic monuments in New England, and join the astonishing quest for lost cities throughout North America. From the war-torn jungles of Guatemala, Nicaragua and Honduras to the deserts, mountains and fields of Mexico, Canada, and the U.S.A., Childress takes the reader in search of sunken ruins, Viking forts, strange tunnel systems, living dinosaurs, early Chinese explorers, and fantastic lost treasure. Packed with both early and current maps, photos and illustrations.
590 PAGES. 6x9 PAPERBACK. ILLUSTRATED. FOOTNOTES. BIBLIOGRAPHY. INDEX. $16.95. CODE: NCA

LOST CITIES & ANCIENT MYSTERIES OF SOUTH AMERICA
by David Hatcher Childress
Rogue adventurer and maverick archaeologist David Hatcher Childress takes the reader on unforgettable journeys deep into deadly jungles, high up on windswept mountains and across scorching deserts in search of lost civilizations and ancient mysteries. Travel with David and explore stone cities high in mountain forests and hear fantastic tales of Inca treasure, living dinosaurs, and a mysterious tunnel system. Whether he is hopping freight trains, searching for secret cities, or just dealing with the daily problems of food, money, and romance, the author keeps the reader spellbound. Includes both early and current maps, photos, and illustrations, and plenty of advice for the explorer planning his or her own journey of discovery.
381 PAGES. 6x9 PAPERBACK. ILLUSTRATED. FOOTNOTES. BIBLIOGRAPHY. INDEX. $16.95. CODE: SAM

LOST CITIES & ANCIENT MYSTERIES OF AFRICA & ARABIA
by David Hatcher Childress
Across ancient deserts, dusty plains and steaming jungles, maverick archaeologist David Childress continues his world-wide quest for lost cities and ancient mysteries. Join him as he discovers forbidden cities in the Empty Quarter of Arabia; "Atlantean" ruins in Egypt and the Kalahari desert; a mysterious, ancient empire in the Sahara; and more. This is the tale of an extraordinary life on the road: across war-torn countries, Childress searches for King Solomon's Mines, living dinosaurs, the Ark of the Covenant and the solutions to some of the fantastic mysteries of the past.
423 PAGES. 6x9 PAPERBACK. ILLUSTRATED. FOOTNOTES & BIBLIOGRAPHY. $14.95. CODE: AFA

24 hour credit card orders—call: 815-253-6390 fax: 815-253-6300
email: auphq@frontiernet.net   www.adventuresunlimitedpress.com   www.wexclub.com

LOST CITIES

TECHNOLOGY OF THE GODS
The Incredible Sciences of the Ancients
by David Hatcher Childress

Popular *Lost Cities* author David Hatcher Childress takes us into the amazing world of ancient technology, from computers in antiquity to the "flying machines of the gods." Childress looks at the technology that was allegedly used in Atlantis and the theory that the Great Pyramid of Egypt was originally a gigantic power station. He examines tales of ancient flight and the technology that it involved; how the ancients used electricity; megalithic building techniques; the use of crystal lenses and the fire from the gods; evidence of various high tech weapons in the past, including atomic weapons; ancient metallurgy and heavy machinery; the role of modern inventors such as Nikola Tesla in bringing ancient technology back into modern use; impossible artifacts; and more.
356 PAGES. 6x9 PAPERBACK. ILLUSTRATED. BIBLIOGRAPHY. $16.95. CODE: TGOD

VIMANA AIRCRAFT OF ANCIENT INDIA & ATLANTIS
by David Hatcher Childress, introduction by Ivan T. Sanderson

Did the ancients have the technology of flight? In this incredible volume on ancient India, authentic Indian texts such as the *Ramayana* and the *Mahabharata* are used to prove that ancient aircraft were in use more than four thousand years ago. Included in this book is the entire Fourth Century BC manuscript *Vimaanika Shastra* by the ancient author Maharishi Bharadwaaja, translated into English by the Mysore Sanskrit professor G.R. Josyer. Also included are chapters on Atlantean technology, the incredible Rama Empire of India and the devastating wars that destroyed it. Also an entire chapter on mercury vortex propulsion and mercury gyros, the power source described in the ancient Indian texts. Not to be missed by those interested in ancient civilizations or the UFO enigma.
334 PAGES. 6x9 PAPERBACK. RARE PHOTOGRAPHS, MAPS AND DRAWINGS. $15.95. CODE: VAA

LOST CONTINENTS & THE HOLLOW EARTH
I Remember Lemuria and the Shaver Mystery
by David Hatcher Childress & Richard Shaver

Lost Continents & the Hollow Earth is Childress' thorough examination of the early hollow earth stories of Richard Shaver and the fascination that fringe fantasy subjects such as lost continents and the hollow earth have had for the American public. Shaver's rare 1948 book *I Remember Lemuria* is reprinted in its entirety, and the book is packed with illustrations from Ray Palmer's *Amazing Stories* magazine of the 1940s. Palmer and Shaver told of tunnels running through the earth—tunnels inhabited by the Deros and Teros, humanoids from an ancient spacefaring race that had inhabited the earth, eventually going underground, hundreds of thousands of years ago. Childress discusses the famous hollow earth books and delves deep into whatever reality may be behind the stories of tunnels in the earth. Operation High Jump to Antarctica in 1947 and Admiral Byrd's bizarre statements, tunnel systems in South America and Tibet, the underground world of Agartha, the belief of UFOs coming from the South Pole, more.
344 PAGES. 6x9 PAPERBACK. ILLUSTRATED. $16.95. CODE: LCHE

A HITCHHIKER'S GUIDE TO ARMAGEDDON
by David Hatcher Childress

With wit and humor, popular Lost Cities author David Hatcher Childress takes us around the world and back in his trippy finalé to the Lost Cities series. He's off on an adventure in search of the apocalypse and end times. Childress hits the road from the fortress of Megiddo, the legendary citadel in northern Israel where Armageddon is prophesied to start. Hitchhiking around the world, Childress takes us from one adventure to another, to ancient cities in the deserts and the legends of worlds before our own. Childress muses on the rise and fall of civilizations, and the forces that have shaped mankind over the millennia, including wars, invasions and cataclysms. He discusses the ancient Armageddons of the past, and chronicles recent Middle East developments and their ominous undertones. In the meantime, he becomes a cargo cult god on a remote island off New Guinea, gets dragged into the Kennedy Assassination by one of the "conspirators," investigates a strange power operating out of the Altai Mountains of Mongolia, and discovers how the Knights Templar and their off-shoots have driven the world toward an epic battle centered around Jerusalem and the Middle East.
320 PAGES. 6x9 PAPERBACK. ILLUSTRATED. BIBLIOGRAPHY. INDEX. $16.95. CODE: HGA

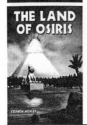

THE LAND OF OSIRIS
An Introduction to Khemitology
by Stephen S. Mehler

Was there an advanced prehistoric civilization in ancient Egypt? Were they the people who built the great pyramids and carved the Great Sphinx? Did the pyramids serve as energy devices and not as tombs for kings? Mehler has uncovered an indigenous oral tradition that still exists in Egypt, and has been fortunate to have studied with a living master of this tradition, Abd'El Hakim Awyan. Mehler has also been given permission to present these teachings to the Western world, teachings that unfold a whole new understanding of ancient Egypt and have only been presented heretofore in fragments by other researchers. Chapters include: Egyptology and Its Paradigms; Khemitology—New Paradigms; Asgat Nefer—The Harmony of Water; Khemit and the Myth of Atlantis; The Extraterrestrial Question; more.
272 PAGES. 6x9 PAPERBACK. ILLUSTRATED. COLOR SECTION. BIBLIOGRAPHY. $18.95. CODE: LOOS

IN QUEST OF LOST WORLDS
Journey to Mysterious Algeria, Ethiopia & the Yucatan
by Count Byron Khun de Prorok

Finally, a reprint of Count Byron de Prorok's classic archeology/adventure book first published in 1936 by E.P. Dutton & Co. in New York. In this exciting and well illustrated book, de Prorok takes us into the deep Sahara of forbidden Algeria, to unknown Ethiopia, and to the many prehistoric ruins of the Yucatan. Includes: Tin Hinan, Legendary Queen of the Tuaregs; The mysterious A'Haggar Range of southern Algeria; Jupiter, Ammon and Tripolitania; The "Talking Dune"; The Land of the Garamantes; Mexico and the Poison Trail; Seeking Atlantis—Chichen Itza; Shadowed by the "Little People"—the Lacandon Pygmie Maya; Ancient Pyramids of the Usamasinta and Piedras Negras in Guatemala; In Search of King Solomon's Mines & the Land of Ophir; Ancient Emerald Mines of Ethiopia. Also included in this book are 24 pages of special illustrations of the famous—and strange—wall paintings of the Ahaggar from the rare book *The Search for the Tassili Frescoes* by Henri Lhote (1959). A visual treat of a remote area of the world that is even today forbidden to outsiders!
324 PAGES. 6x9 PAPERBACK. ILLUSTRATED. $16.95. CODE: IQLW

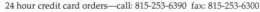

24 hour credit card orders—call: 815-253-6390 fax: 815-253-6300
email: auphq@frontiernet.net   www.adventuresunlimitedpress.com   www.wexclub.com

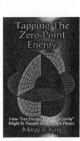

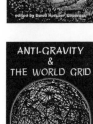

FREE ENERGY SYSTEMS

LOST SCIENCE
by Gerry Vassilatos

Rediscover the legendary names of suppressed scientific revolution—remarkable lives, astounding discoveries, and incredible inventions which would have produced a world of wonder. How did the aura research of Baron Karl von Reichenbach prove the vitalistic theory and frighten the greatest minds of Germany? How did the physiophone and wireless of Antonio Meucci predate both Bell and Marconi by decades? How does the earth battery technology of Nathan Stubblefield portend an unsuspected energy revolution? How did the geoaetheric engines of Nikola Tesla threaten the establishment of a fuel-dependent America? The microscopes and virus-destroying ray machines of Dr. Royal Rife provided the solution for every world-threatening disease. Why did the FDA and AMA together condemn this great man to Federal Prison? The static crashes on telephone lines enabled Dr. T. Henry Moray to discover the reality of radiant space energy. Was the mysterious "Swedish stone," the powerful mineral which Dr. Moray discovered, the very first historical instance in which stellar power was recognized and secured on earth? Why did the Air Force initially fund the gravitational warp research and warp-cloaking devices of T. Townsend Brown and then reject it? When the controlled fusion devices of Philo Farnsworth achieved the "break-even" point in 1967 in the FUSOR project was abruptly cancelled by ITT.
304 PAGES. 6x9 PAPERBACK. ILLUSTRATED. BIBLIOGRAPHY. $16.95. CODE: LOS

SECRETS OF COLD WAR TECHNOLOGY
Project HAARP and Beyond
by Gerry Vassilatos

Vassilatos reveals that "Death Ray" technology has been secretly researched and developed since the turn of the century. Included are chapters on such inventors and their devices as H.C. Vion, the developer of auroral energy receivers; Dr. Selim Lemstrom's pre-Tesla experiments; the early beam weapons of Grindell-Mathews, Ulivi, Turpain and others; John Hettenger and his early beam power systems. Learn about Project Argus, Project Teak and Project Orange; EMP experiments in the 60s; why the Air Force directed the construction of a huge Ionospheric "backscatter" telemetry system across the Pacific just after WWII; why Raytheon has collected every patent relevant to HAARP over the past few years; more.
250 PAGES. 6x9 PAPERBACK. ILLUSTRATED. $15.95. CODE: SCWT

QUEST FOR ZERO-POINT ENERGY
Engineering Principles for "Free Energy"
by Moray B. King

King expands, with diagrams, on how free energy and anti-gravity are possible. The theories of zero point energy maintain there are tremendous fluctuations of electrical field energy embedded within the fabric of space. King explains the following topics: Tapping the Zero-Point Energy as an Energy Source; Fundamentals of a Zero-Point Energy Technology; Vacuum Energy Vortices; The Super Tube; Charge Clusters: The Basis of Zero-Point Energy Inventions; Vortex Filaments, Torsion Fields and the Zero-Point Energy; Transforming the Planet with a Zero-Point Energy Experiment; Dual Vortex Forms: The Key to a Large Zero-Point Energy Coherence. Packed with diagrams, patents and photos. With power shortages now a daily reality in many parts of the world, this book offers a fresh approach very rarely mentioned in the mainstream media.
224 PAGES. 6x9 PAPERBACK. ILLUSTRATED. $14.95. CODE: QZPE

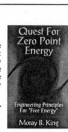

Quest For Zero Point Energy

Engineering Principles For "Free Energy"

Moray B. King

THE TIME TRAVEL HANDBOOK
A Manual of Practical Teleportation & Time Travel
edited by David Hatcher Childress

In the tradition of *The Anti-Gravity Handbook* and *The Free-Energy Device Handbook*, science and UFO author David Hatcher Childress takes us into the weird world of time travel and teleportation. Not just a whacked-out look at science fiction, this book is an authoritative chronicling of real-life time travel experiments, teleportation devices and more. *The Time Travel Handbook* takes the reader beyond the government experiments and deep into the uncharted territory of early time travellers such as Nikola Tesla and Guglielmo Marconi and their alleged time travel experiments, as well as the Wilson Brothers of EMI and their connection to the Philadelphia Experiment—the U.S. Navy's forays into invisibility, time travel, and teleportation. Childress looks into the claims of time travelling individuals, and investigates the unusual claim that the pyramids on Mars were built in the future and sent back in time. A highly visual, large format book, with patents, photos and schematics. Be the first on your block to build your own time travel device!
316 PAGES. 7x10 PAPERBACK. ILLUSTRATED. $16.95. CODE: TTH

THE TESLA PAPERS
Nikola Tesla on Free Energy & Wireless Transmission of Power
by Nikola Tesla, edited by David Hatcher Childress

David Hatcher Childress takes us into the incredible world of Nikola Tesla and his amazing inventions. Tesla's rare article "The Problem of Increasing Human Energy with Special Reference to the Harnessing of the Sun's Energy" is included. This lengthy article was originally published in the June 1900 issue of *The Century Illustrated Monthly Magazine* and it was the outline for Tesla's master blueprint for the world. Tesla's fantastic vision of the future, including wireless power, anti-gravity, free energy and highly advanced solar power. Also included are some of the papers, patents and material collected on Tesla at the Colorado Springs Tesla Symposiums, including papers on: •The Secret History of Wireless Transmission •Tesla and the Magnifying Transmitter •Design and Construction of a Half-Wave Tesla Coil •Electrostatics: A Key to Free Energy •Progress in Zero-Point Energy Research •Electromagnetic Energy from Antennas to Atoms •Tesla's Particle Beam Technology •Fundamental Excitatory Modes of the Earth-Ionosphere Cavity
325 PAGES. 8X10 PAPERBACK. ILLUSTRATED. $16.95. CODE: TTP

THE FANTASTIC INVENTIONS OF NIKOLA TESLA
by Nikola Tesla with additional material by David Hatcher Childress

This book is a readable compendium of patents, diagrams, photos and explanations of the many incredible inventions of the originator of the modern era of electrification. In Tesla's own words are such topics as wireless transmission of power, death rays, and radio-controlled airships. In addition, rare material on German bases in Antarctica and South America, and a secret city built at a remote jungle site in South America by one of Tesla's students, Guglielmo Marconi. Marconi's secret group claims to have built flying saucers in the 1940s and to have gone to Mars in the early 1950s! Incredible photos of these Tesla craft are included. The Ancient Atlantean system of broadcasting energy through a grid system of obelisks and pyramids is discussed, and a fascinating concept comes out of one chapter: that Egyptian engineers had to wear protective metal head-shields while in these power plants, hence the Egyptian Pharoah's head covering as well as the Face on Mars! •His plan to transmit free electricity into the atmosphere. •How electrical devices would work using only small antennas. •Why unlimited power could be utilized anywhere on earth. •How radio and radar technology can be used as death-ray weapons in Star Wars.
342 PAGES. 6x9 PAPERBACK. ILLUSTRATED. $16.95. CODE: FINT

24 hour credit card orders—call: 815-253-6390 fax: 815-253-6300

email: auphq@frontiernet.net   www.adventuresunlimitedpress.com   www.wexclub.com

HISTORY—CONSPIRACY

POPULAR PARANOIA
The Best of Steamshovel Press
edited by Kenn Thomas

The anthology exposes the biologocal warfare origins of AIDS; the Nazi/Nation of Islam link; the cult of Elizabeth Clare Prophet; the Oklahoma City bombing writings of the late Jim Keith, as well as an article on Keith's own strange death; the conspiratorial mind of John Judge; Marion Pettie and the shadowy Finders group in Washington, DC; demonic iconography; the death of Princess Diana, its connection to the Octopus and the Saudi aerospace contracts; spies among the Rajneeshis; scholarship on the historic Illuminati; and many other parapolitical topics. The book also includes the Steamshovel's last-ever interviews with the great Beat writers Allen Ginsberg and William S. Burroughs, and neuronaut Timothy Leary, and new views of the master Beat, Neal Cassady and Jack Kerouac's science fiction.

308 PAGES. 8X10 PAPERBACK. ILLUSTRATED. $19.95. CODE: POPA

THE ORION PROPHECY
Egyptian and Mayan Prophecies on the Cataclysm of 2012
by Patrick Geryl and Gino Ratinckx

In the year 2012 the Earth awaits a super catastrophe: its magnetic field will reverse in one go. Phenomenal earthquakes and tidal waves will completely destroy our civilization. Europe and North America will shift thousands of kilometers northwards into polar climes. Nearly everyone will perish in the apocalyptic happenings. These dire predictions stem from the Mayans and Egyptians—descendants of the legendary Atlantis. The Atlanteans had highly evolved astronomical knowledge and were able to exactly predict the previous world-wide flood in 9792 BC. They built tens of thousands of boats and escaped to South America and Egypt. In the year 2012 Venus, Orion and several others stars will take the same 'code-positions' as in 9792 BC! For thousands of years historical sources have told of a forgotten time capsule of ancient wisdom located in a labyrinth of secret chambers filled with artifacts and documents from the previous flood. We desperately need this information now—and this book gives one possible location.

324 PAGES. 6X9 PAPERBACK. ILLUSTRATED. BIBLIOGRAPHY. $16.95. CODE: ORP

THE SHADOW GOVERNMENT
9-11 and State Terror
by Len Bracken, introduction by Kenn Thomas

Bracken presents the alarming yet convincing theory that nation-states engage in or allow terror to be visited upon their citizens. It is not just liberation movements and radical groups that deploy terroristic tactics for offensive ends. States use terror defensively to directly intimidate their citizens and to indirectly attack themselves or harm their citizens under a false flag. Their motives? To provide pretexts for war or for increased police powers or both. This stratagem of indirectly using terrorism has been executed by statesmen in various ways but tends to involve the pretense of blind eyes, misdirection, and cover-ups that give statesmen plausible deniability. Lusitiania, Pearl Harbor, October Surprise, the first World Trade Center bombing, the Oklahoma City bombing and other well-known incidents suggest that terrorism is often and successfully used by states in an indirectly defensive way to take the offensive against enemies at home and abroad. Was 9-11 such an indirect defensive attack?

288 PAGES. 6X9 PAPERBACK. ILLUSTRATED. $16.00. CODE: SGOV

MASS CONTROL
Engineering Human Consciousness
by Jim Keith

Conspiracy expert Keith's final book on mind control, Project Monarch, and mass manipulation presents chilling evidence that we are indeed spinning a Matrix. Keith describes the New Man, whose conception of reality is a dance of electronic images fired into his forebrain, a gossamer construction of his masters, designed so that he will not—under any circumstances—perceive the actual. His happiness is delivered to him through a tube or an electronic connection. His God lurks behind an electronic curtain; when the curtain is pulled away we find the CIA sorcerer, the media manipulator... Chapters on the CIA, Tavistock, Jolly West and the Violence Center, Guerrilla Mindwar, Brice Taylor, other recent "victims," more.

256 PAGES. 6X9 PAPERBACK. ILLUSTRATED. INDEX. $16.95. CODE: MASC

WAKE UP DOWN THERE!
The Excluded Middle Anthology
by Greg Bishop

The great American tradition of dropout culture makes it over the millennium mark with a collection of the best from *The Excluded Middle*, the critically acclaimed underground zine of UFOs, the paranormal, conspiracies, psychedelia, and spirit. Contributions from Robert Anton Wilson, Ivan Stang, Martin Kottmeyer, John Shirley, Scott Corrales, Adam Gorightly and Robert Sterling; and interviews with James Moseley, Karla Turner, Bill Moore, Kenn Thomas, Richard Boylan, Dean Radin, Joe McMoneagle, and the mysterious Ira Einhorn (an *Excluded Middle* exclusive). Includes full versions of interviews and extra material not found in the newsstand versions.

420 PAGES. 8X11 PAPERBACK. ILLUSTRATED. $25.00. CODE: WUDT

DARK MOON
Apollo and the Whistleblowers
by Mary Bennett and David Percy

•Was Neil Armstrong really the first man on the Moon?
•Did you know a second craft was going to the Moon at the same time as Apollo 11?
•Do you know that potentially lethal radiation is prevalent throughout deep space?
•Do you know there are serious discrepancies in the account of the Apollo 13 'accident'?
•Did you know that 'live' color TV from the Moon was not actually live at all?
•Did you know that the Lunar Surface Camera had no viewfinder?
•Do you know that lighting was used in the Apollo photographs—yet no lighting equipment was taken to the Moon?

All these questions, and more, are discussed in great detail by British researchers Bennett and Percy in *Dark Moon*, the definitive book (nearly 600 pages) on the possible faking of the Apollo Moon missions. Bennett and Percy delve into every possible aspect of this beguiling theory, one that rocks the very foundation of our beliefs concerning NASA and the space program. Tons of NASA photos analyzed for possible deceptions.

568 PAGES. 6X9 PAPERBACK. ILLUSTRATED. BIBLIOGRAPHY. INDEX. $25.00. CODE: DMO

24 hour credit card orders—call: 815-253-6390 fax: 815-253-6300

email: auphq@frontiernet.net   www.adventuresunlimitedpress.com   www.wexclub.com

MYSTIC TRAVELLER SERIES

THE MYSTERY OF EASTER ISLAND
by Katherine Routledge
The reprint of Katherine Routledge's classic archaeology book which was first published in London in 1919. The book details her journey by yacht from England to South America, around Patagonia to Chile and on to Easter Island. Routledge explored the amazing island and produced one of the first-ever accounts of the life, history and legends of this strange and remote place. Routledge discusses the statues, pyramid-platforms, Rongo Rongo script, the Bird Cult, the war between the Short Ears and the Long Ears, the secret caves, ancient roads on the island, and more. This rare book serves as a sourcebook on the early discoveries and theories on Easter Island.
432 PAGES. 6X9 ILLUSTRATED. $16.95. CODE: MEI

THIS RARE ARCHAEOLOGY BOOK ON
EASTER ISLAND IS BACK IN PRINT!

MYSTERY CITIES
OF THE MAYA
by Thomas Gann

MYSTERY CITIES OF THE MAYA
Exploration and Adventure in Lubaantun & Belize
by Thomas Gann
First published in 1925, *Mystery Cities of the Maya* is a classic in Central American archaeology-adventure. Gann was close friends with Mike Mitchell-Hedges, the British adventurer who discovered the famous crystal skull with his adopted daughter Sammy and Lady Richmond Brown, their benefactress. Gann battles pirates along Belize's coast and goes upriver with Mitchell-Hedges to the site of Lubaantun where they excavate a strange lost city where the crystal skull was discovered. Lubaantun is a unique city in the Mayan world as it is built out of precisely carved blocks of stone without the usual plaster-cement facing. Lubaantun contained several large pyramids partially destroyed by earthquakes and a large amount of artifacts. Gann shared Mitchell-Hedges belief in Atlantis and lost civilizations (pre-Mayan) in Central America and the Caribbean. Lots of good photos, maps and diagrams.
252 PAGES. 6X9 PAPERBACK. ILLUSTRATED. $16.95. CODE: MCOM

IN SECRET TIBET
T. Illion

IN SECRET TIBET
by Theodore Illion
Reprint of a rare 30s adventure travel book. Illion was a German wayfarer who not only spoke fluent Tibetan, but travelled in disguise as a native through forbidden Tibet when it was off-limits to all outsiders. His incredible adventures make this one of the most exciting travel books ever published. Includes illustrations of Tibetan monks levitating stones by acoustics.
210 PAGES. 6X9 PAPERBACK. ILLUSTRATED. $15.95. CODE: IST

MYSTIC TRAVELLER SERIES

DARKNESS OVER TIBET
by Theodore Illion
In this second reprint of Illion's rare books, the German traveller continues his journey through Tibet and is given directions to a strange underground city. As the original publisher's remarks said, "this is a rare account of an underground city in Tibet by the only Westerner ever to enter it and escape alive! "
210 PAGES. 6X9 PAPERBACK. ILLUSTRATED. $15.95. CODE: DOT

Danger My Ally

DANGER MY ALLY
The Amazing Life Story of the Discoverer of the Crystal Skull
by "Mike" Mitchell-Hedges
The incredible life story of "Mike" Mitchell-Hedges, the British adventurer who discovered the Crystal Skull in the lost Mayan city of Lubaantun in Belize. Mitchell-Hedges has lived an exciting life: gambling everything on a trip to the Americas as a young man, riding with Pancho Villa, questing for Atlantis, fighting bandits in the Caribbean and discovering the famous Crystal Skull.
374 PAGES. 6X9 PAPERBACK. ILLUSTRATED. BIBLIOGRAPHY & INDEX. $16.95. CODE: DMA

The true life adventure of
F.A. Mitchell-Hedges

IN SECRET MONGOLIA
by Henning Haslund
First published by Kegan Paul of London in 1934, Haslund takes us into the barely known world of Mongolia of 1921, a land of god-kings, bandits, vast mountain wilderness and a Russian army running amok. Starting in Peking, Haslund journeys to Mongolia as part of the Krebs Expedition—a mission to establish a Danish butter farm in a remote corner of northern Mongolia. Along the way, he smuggles guns and nitroglycerin, is thrown into a prison by the new Communist regime, battles the Robber Princess and more. In Mongolia we meet the "Mad Baron" Ungern-Sternberg and his renegade Russian army, the many characters of Urga's fledgling foreign community, and the last god-king of Mongolia, Seng Chen Gegen, the fifth reincarnation of the Tiger god and the "ruler of all Torguts." Aside from the esoteric and mystical material, there is plenty of just plain adventure: Haslund encounters a Mongolian werewolf; is ambushed along the trail; escapes from prison and fights terrifying blizzards; more.
374 PAGES. 6X9 PAPERBACK. ILLUSTRATED. BIBLIOGRAPHY & INDEX. $16.95. CODE: ISM

MEN & GODS IN MONGOLIA
by Henning Haslund
First published in 1935 by Kegan Paul of London, Haslund takes us to the lost city of Karakota in the Gobi desert. We meet the Bodgo Gegen, a god-king in Mongolia similar to the Dalai Lama of Tibet. We meet Dambin Jansang, the dreaded warlord of the "Black Gobi." There is even material in this incredible book on the Hi-mori, an "airhorse" that flies through the sky (similar to a Vimana) and carries with it the sacred stone of Chintamani. Aside from the esoteric and mystical material, there is plenty of just plain adventure: Haslund and companions journey across the Gobi desert by camel caravan; are kidnapped and held for ransom; witness initiation into Shamanic societies; meet reincarnated warlords; and experience the violent birth of "modern" Mongolia.
358 PAGES. 6X9 PAPERBACK. ILLUSTRATED. INDEX. $15.95. CODE: MGM

This rare 1935 book is back in print!
Mystic Traveller Series

24 hour credit card orders—call: 815-253-6390 fax: 815-253-6300
email: auphq@frontiernet.net   www.adventuresunlimitedpress.com   www.wexclub.com

ATLANTIS REPRINT SERIES

ATLANTIS: MOTHER OF EMPIRES
Atlantis Reprint Series
by Robert Stacy-Judd

Robert Stacy-Judd's classic 1939 book on Atlantis is back in print in this large-format paperback edition. Stacy-Judd was a California architect and an expert on the Mayas and their relationship to Atlantis. He was an excellent artist and his work is lavishly illustrated. The eighteen comprehensive chapters in the book are: The Mayas and the Lost Atlantis; Conjectures and Opinions; The Atlantean Theory; Cro-Magnon Man; East is West; And West is East; The Mormons and the Mayas; Astrology in Two Hemispheres; The Language of Architecture; The American Indian; Pre-Panamanians and Pre-Incas; Columns and City Planning; Comparisons and Mayan Art; The Iberian Link; The Maya Tongue; Quetzalcoatl; Summing Up the Evidence; The Mayas in Yucatan.
340 PAGES. 8x11 PAPERBACK. ILLUSTRATED. INDEX. $19.95. CODE: AMOE

MYSTERIES OF ANCIENT SOUTH AMERICA
Atlantis Reprint Series
by Harold T. Wilkins

The reprint of Wilkins' classic book on the megaliths and mysteries of South America. This book predates Wilkin's book *Secret Cities of Old South America* published in 1952. *Mysteries of Ancient South America* was first published in 1947 and is considered a classic book of its kind. With diagrams, photos and maps, Wilkins digs into old manuscripts and books to bring us some truly amazing stories of South America: a bizarre subterranean tunnel system; lost cities in the remote border jungles of Brazil; legends of Atlantis in South America; cataclysmic changes that shaped South America; and other strange stories from one of the world's great researchers. Chapters include: Our Earth's Greatest Disaster, Dead Cities of Ancient Brazil, The Jungle Light that Shines by Itself, The Missionary Men in Black: Forerunners of the Great Catastrophe, The Sign of the Sun: The World's Oldest Alphabet, Sign-Posts to the Shadow of Atlantis, The Atlanean "Subterraneans" of the Incas, Tiahuanacu and the Giants, more.
236 PAGES. 6x9 PAPERBACK. ILLUSTRATED. INDEX. $14.95. CODE: MASA

SECRET CITIES OF OLD SOUTH AMERICA
Atlantis Reprint Series
by Harold T. Wilkins

The reprint of Wilkins' classic book, first published in 1952, claiming that South America was Atlantis. Chapters include Mysteries of a Lost World; Atlantis Unveiled; Red Riddles on the Rocks; South America's Amazons Existed!; The Mystery of El Dorado and Gran Payatiti—the Final Refuge of the Incas; Monstrous Beasts of the Unexplored Swamps & Wilds; Weird Denizens of Antediluvian Forests; New Light on Atlantis from the World's Oldest Book; The Mystery of Old Man Noah and the Arks; and more.
438 PAGES. 6x9 PAPERBACK. ILLUSTRATED. BIBLIOGRAPHY & INDEX. $16.95. CODE: SCOS

THE SHADOW OF ATLANTIS
The Echoes of Atlantean Civilization Tracked through Space & Time
by Colonel Alexander Braghine

First published in 1940, *The Shadow of Atlantis* is one of the great classics of Atlantis research. The book amasses a great deal of archaeological, anthropological, historical and scientific evidence in support of a lost continent in the Atlantic Ocean. Braghine covers such diverse topics as Egyptians in Central America, the myth of Quetzalcoatl, the Basque language and its connection with Atlantis, the connections with the ancient pyramids of Mexico, Egypt and Atlantis, the sudden demise of mammoths, legends of giants and much more. Braghine was a linguist and spends part of the book tracing ancient languages to Atlantis and studying little-known inscriptions in Brazil, deluge myths and the connections between ancient languages. Braghine takes us on a fascinating journey through space and time in search of the lost continent.
288 PAGES. 6x9 PAPERBACK. ILLUSTRATED. $16.95. CODE: SOA

RIDDLE OF THE PACIFIC
by John Macmillan Brown

Oxford scholar Brown's classic work on lost civilizations of the Pacific is now back in print! John Macmillan Brown was an historian and New Zealand's premier scientist when he wrote about the origins of the Maoris. After many years of travel thoughout the Pacific studying the people and customs of the south seas islands, he wrote *Riddle of the Pacific* in 1924. The book is packed with rare turn-of-the-century illustrations. Don't miss Brown's classic study of Easter Island, ancient scripts, megalithic roads and cities, more. Brown was an early believer in a lost continent in the Pacific.
460 PAGES. 6x9 PAPERBACK. ILLUSTRATED. $16.95. CODE: ROP

THE HISTORY OF ATLANTIS
by Lewis Spence

Lewis Spence's classic book on Atlantis is now back in print! Spence was a Scottish historian (1874-1955) who is best known for his volumes on world mythology and his five Atlantis books. *The History of Atlantis* (1926) is considered his finest. Spence does his scholarly best in chapters on the Sources of Atlantean History, the Geography of Atlantis, the Races of Atlantis, the Kings of Atlantis, the Religion of Atlantis, the Colonies of Atlantis, more. Sixteen chapters in all.
240 PAGES. 6x9 PAPERBACK. ILLUSTRATED WITH MAPS, PHOTOS & DIAGRAMS. $16.95. CODE: HOA

ATLANTIS IN SPAIN
A Study of the Ancient Sun Kingdoms of Spain
by E.M. Whishaw

First published by Rider & Co. of London in 1928, this classic book is a study of the megaliths of Spain, ancient writing, cyclopean walls, sun worshipping empires, hydraulic engineering, and sunken cities. An extremely rare book, it was out of print for 60 years. Learn about the Biblical Tartessus; an Atlantean city at Niebla; the Temple of Hercules and the Sun Temple of Seville; Libyans and the Copper Age; more. Profusely illustrated with photos, maps and drawings.
284 PAGES. 6x9 PAPERBACK. ILLUSTRATED. TABLES OF ANCIENT SCRIPTS. $15.95. CODE: AIS

One Adventure Place
P.O. Box 74
Kempton, Illinois 60946
United States of America
•Tel.: 1-800-718-4514 or 815-253-6390
•Fax: 815-253-6300
Email: auphq@frontiernet.net
http://www.adventuresunlimitedpress.com
or www.adventuresunlimited.nl

10% Discount when you order 3 or more items!

ORDERING INSTRUCTIONS

✓ Remit by USD$ Check, Money Order or Credit Card
✓ Visa, Master Card, Discover & AmEx Accepted
✓ Prices May Change Without Notice
✓ 10% Discount for 3 or more Items

SHIPPING CHARGES

United States

✓ Postal Book Rate { $3.00 First Item
50¢ Each Additional Item

✓ Priority Mail { $4.50 First Item
$2.00 Each Additional Item

✓ UPS { $5.00 First Item
$1.50 Each Additional Item

NOTE: UPS Delivery Available to Mainland USA Only

Canada

✓ Postal Book Rate { $6.00 First Item
$2.00 Each Additional Item

✓ Postal Air Mail { $8.00 First Item
$2.50 Each Additional Item

✓ Personal Checks or Bank Drafts MUST BE
✓ USD$ and Drawn on a US Bank
Canadian Postal Money Orders in US$ OK
✓ Payment MUST BE US$

All Other Countries

✓ Surface Delivery { $10.00 First Item
$4.00 Each Additional Item

✓ Postal Air Mail { $14.00 First Item
$5.00 Each Additional Item

✓ Checks and Money Orders MUST BE US$
and Drawn on a US Bank or branch.

✓ Payment by credit card preferred!

SPECIAL NOTES

✓ RETAILERS: Standard Discounts Available
✓ BACKORDERS: We Backorder all Out-of-
Stock Items Unless Otherwise Requested
✓ PRO FORMA INVOICES: Available on Request
✓ VIDEOS: NTSC Mode Only. Replacement only.
✓ For PAL mode videos contact our other offices:

European Office:
Adventures Unlimited, Pannewal 22,
Enkhuizen, 1602 KS, The Netherlands
http: www.adventuresunlimited.nl
Check Us Out Online at:
www.adventuresunlimitedpress.com

Please check: ☑

☐ This is my first order	☐ I have ordered before	☐ This is a new address

Name	
Address	
City	
State/Province	Postal Code
Country	
Phone day	Evening
Fax	Email

Item Code	Item Description	Price	Qty	Total

Please check: ☑

☐ Postal-Surface

☐ Postal-Air Mail
(Priority in USA)

☐ UPS
(Mainland USA only)

Subtotal ➡	
Less Discount-10% for 3 or more items ➡	
Balance ➡	
Illinois Residents 6.25% Sales Tax ➡	
Previous Credit ➡	
Shipping ➡	
Total (check/MO in USD$ only) ➡	

☐ Visa/MasterCard/Discover/Amex

Card Number

Expiration Date

10% Discount When You Order 3 or More Items!

Comments & Suggestions	Share Our Catalog with a Friend